Hazardous Building Materials

Second Edition

Hazardous Building Materials

A guide to the selection of environmentally responsible alternatives

/4010/

Steve Curwell, Bob Fox, Morris Greenberg and Chris March

London and New York

First published 2002 by Spon Press
11 New Fetter Lane, London EC4P 4EE

Simultaneously published in the USA and Canada
by Spon Press
29 West 35th Street, New York, NY 10001

Spon Press is an imprint of the Taylor & Francis Group

© 2002 BCF Publishing Ltd

Printed and bound in Great Britain by TJ International Ltd, Padstow, Cornwall.

Publishers Note
This book has been produced from camera-ready copy supplied by the authors.

British Library Cataloguing in Publication Data
A catalogue record for this book is available from the British Library

Library of Congress Cataloging in Publication Data
A catalog record for this book has been requested

ISBN 0-419-23450-0

CONTENTS

Authors

Steve Curwell BSc RIBA is a graduate of Queens University, Belfast with broad experience of environmental and sustainable development research. He is currently senior lecturer in Environmental Technology at the University of Salford and director of a major E.U. research network called BEQUEST exploring sustainable urban development. He is a member of the Sustainable Construction Working Group of the UK Construction Research and Information Strategy Panel and of the Conseil International du Batimént Working Group 100 – Environmental Assessment of Buildings. Steve was chairman of the Green Building Challenge UK selection panel. He was co-author and editor of the first edition of *Hazardous Building Materials*, co-authored of *CFCs in Buildings*, and co-editor and contributor to *Buildings and Health – the Rosehaugh Guide*.

Bob Fox BSc ARICS is an Associate with EC Harris Capital Project and Facilities Consultants. He is a Chartered Quantity Surveyor and has, since the early 1980's, specialised in the field of buildings and health. Working in both the public and private sectors, he was a contributor to the first edition of *Hazardous Building Materials: A Guide to the Selection of Alternatives*, co-author of *CFCs in Buildings* and has lectured and written on the subject of costs associated with buildings, health and environmental issues.

Morris Greenberg MB FRCP FFOM. Graduate of University College Hospital Medical School, he practiced internal medicine before joining Philips Industries. Transferred to HM Inspectorate of Factories and was appointed Honorary Lecturer, the Cardiothoracic Institute. Subsequently Senior Medical Officer, the Division of Toxicology and Environmental Health, the Department of Health. Acted as consultant to, the European Commission, WHO, and ILO, and has been visiting Professor at Tulane University School of Medicine and at Mount Sinai School of Medicine, New York. He has served on various BSI committees, and has acted as Medical Secretary to expert committees. He contributed to the first edition of *Hazardous Building Materials: A Guide to the Selection of Alternatives* and *Buildings and Health: The Rosehaugh Guide* and has published on occupational morbidity, mortality and cancer registration.

Chris March BSc (Tech) MCIOB graduated with a degree in Building from Manchester University in 1963. As a Chartered Builder, he worked for John Laing Construction and later for John Laing Concrete where he became Factory Manager. In 1971 entered higher education, working in both the UK and Hong Kong before joining the University of Salford, now as Senior Lecturer. He was Dean of the Faculty of the Environment from 1996-1999. In 1986 he won the Council for Higher Education Construction Industry Partnership Award for Innovation. He was co-author and editor of *Hazardous Building Materials: A Guide to the Selection of Alternatives*, co-author of *CFCs in Building* and co-editor and contributor to *Buildings and Health: The Rosehaugh Guide*. He has written widely on buildings and health and is a consultant in this field and the environment.

Acknowledgements

This research project, which has culminated in the publication of this guide, occurred as a result of Dr Godfrey Bradman's concern about environmental pollution and the health and general well being of our society and we gratefully acknowledge his initiative, his support for the project and his encouragement to all those involved.

FURTHER ACKNOWLEDGEMENTS

We would like to record our contribution and our thanks to:

Mike Anderson, Managing Director of LAR Ltd, for his contribution on matters associated with the removal of asbestos from existing buildings.

David Dowdle, Lecturer in the School of Construction and Property Management, the University of Salford, for his contribution on building services in Chapter 4.

Norma Ford, Senior Lecturer in the School of Environment and Life Sciences, the University of Salford for her contribution on matters pertaining to lead in existing buildings.

Samantha Nicholson, consultant to Groundwork Trust Salford and Trafford, for her contribution on general environmental issues and sustainable development in Chapter 3.

Finally our thanks to all those manufacturers who willingly provided information on their products.

Foreword

It gives me great pleasure to introduce the second edition of Hazardous Building Materials. In 1985 in the preface of the first edition of this book I stated that 'I hope other organisations providing buildings will share my concern. Ignorance may be an excuse for what has happened in the past. This book, and the work that I hope will follow, should make it an unacceptable defence from now on.' Sixteen years later what has changed?

On the positive side, rules such as the Control of Substances Hazardous to Health Regulations have been enacted. This ensures, not just that manufacturers provide information about the hazards associated with the use of their products, but also places obligations on those using these materials to seek out this data and to act upon it. Unfortunately much of this information languishes on the shelf, with little or no attempt made to substitute safer materials.

Trade literature continues to state only the best features of the particular product it serves to promote, with scant regard to potential hazards, leaving the specifier either to miss the problem through ignorance or to spend considerable time and energy seeking better information to resolve the problem. Some manufacturers have now begun to take environmental issues more seriously, perhaps due to uncomfortable problems such as with those experienced with CFCs, or because of perceived commercial advantage over competitors. Often this advantage is portrayed in a carefully designed narrow manner concentrating only on the positive elements.

There are still no laid down maximum exposure limits for members of the public in the household indoor environment for a number of common pollutants of indoor air, other than rough and ready calculations based upon those laid down for the workplace – even where any level of pollution can be tolerated. This is unsatisfactory, and does not take account of vulnerable groups such as our children, the elderly and the housebound, nor of variables such as different ventilation rates, temperature and humidity.

It is alarming that so little research work has been undertaken fully to establish the potential risks from building materials to the health of our population and especially our children. Too many council houses and schools still contain asbestos, and 30% of all buildings are affected by sick building syndrome. More than a third of people are expected to contract cancer in their lifetime and the distinguished epidemiologist Professor Sir Richard Doll now believes that more than 90% of cancers are avoidable. It is surprising how few of the 60,000 chemicals that are used commercially and have been listed by the US authorities under the Toxic Substances Control Act have been identified as potential human carcinogens, and more research urgently needs to be undertaken by Central Government.

Greater protection is therefore needed for occupants of homes and workplaces. Inadequate care is taken by those engaged in the design, specification and construction of buildings and in their operation. Unless the authorities are prepared to enforce more effectively adequate safety codes one can hardly expect circumstances materially to improve.

I commissioned this work to update the first edition and also to bring together the knowledge available on wider environmental issues related to the selection of materials for buildings. My initial interest had been in the desire to provide safe and healthy environments in which people live and work. However, in recent years, in the knowledge that we must use the planet's resources in a more sustainable fashion so as to preserve life as we know it, I have become convinced of the need to include environmental issues in this edition.

I am pleased to share this book with all those who are building the next generation of homes. It is my perception that we have moved only a short distance since 1985. I hope that those involved in the production of homes, in particular those engaged in the manufacture and specification of materials, the development and design of property, and the enactment of legislation, will take up these issues and provide the skill and care needed to ensure our children have a healthy and sustainable environment in which to live. We all share that responsibility.

Godfrey Bradman
London 2001

Preface

This book is provides concise guidance on the selection of building materials and components for dwellings on the basis of technical, health, environmental and cost criteria. Due to the juxtaposition of this information in a data sheet format it enables selection to be made on a much more complete range of factors. The authors believe this to be the most comprehensive approach to these questions available to date.

The first edition of the guide concentrated on the deleterious effects which materials used in the construction of low-rise residential buildings might have on the health of users and occupiers. The Concise Oxford Dictionary defines deleterious as 'being harmful to mind or body', a clear and precise definition particularly appropriate when applied to some of the materials being used at that time in the construction of buildings, such as asbestos and lead. The relative risk to the occupant's health depends upon individual applications of these materials together with environmental factors within the building, notably ventilation. Substitutes are available, but unfortunately these materials may also have some risk attached to their use. Since 1986 considerable more information has become available on the hazards to health and the environment, hence this new edition.

At the time of publication of the first edition a limitation of the text was that the applications selected were either, a known deleterious material or a material that there was some doubt about at the time. The second edition has addressed this issue by increasing the number of data sheets to take account of the majority of applications, although minor applications such as nails, brackets and ironmongery have been omitted. Generic material sheets of the most common building materials have also been produced to support the application sheets. (Chapter 6)

Further, since the publication of the first edition there has been an increasing interest and need to select materials taking into account their environmental impact and wider issue of sustainability. With this in mind, the authors have introduced an environmental assessment in the application sheets (Chapter 6) and an exploration of the wider environmental debate. (Chapters 3 and 4)

The focus is the hazards and risks to the occupiers and the general environment. As in the first edition problems arising in existing buildings and from maintenance and alterations carried out by recognised tradesmen and the DIY occupier are also taken into account. (Chapter 7). In this respect there is some overlap with the requirements of the health and safety of workers and operatives, however this specific dimension has not been explored in detail by the authors.

The problem for the building designer is to select materials which offer the least hazard to health, have as little negative impact on the environment as possible, but are technically and aesthetically satisfactory and remain within sensible cost limits. This new edition is written to assist readers in this task.

Steve Curwell and Chris March 2001

CHAPTER ONE

Introduction

S.R.CURWELL, BSc MSc ARIBA
C.G.MARCH, BSc(Tech) MCIOB

The last half-century has seen a revolution in science and medicine leading to a much clearer understanding of nature and of the human condition and of the natural environmental systems. Now the effects of environmental factors on human life and of human actions on the environment as well as the relationship to health and well being are much more clearly appreciated. The changes of lifestyle and consumption necessary to preserve and maintain the natural systems for the general health and wellbeing of mankind are now much clearer.
This book is concerned with the actions that building designers, specifiers and the others involved in decision making over the nature of buildings should take to protect the health of the users and occupants. This can be addressed directly by reducing hazards to the health of occupants of buildings and indirectly by reducing environmental impacts in the general environment. The latter can be seen in global and local terms. Global issues such as global warming affect the whole planet and are of importance to the health of the global community. Local issues such as the pollution of rivers are of importance to individual nations or to the health of the local community.

1.1 HEALTH ISSUES

The main concern in the recent past has concentrated upon the occupational exposure of the workforce involved in manufacture and processing industries, which has led to a general improvement in the workplace. Concern shown over deleterious materials at the workplace has, quite rightly, developed an increasing awareness of the problems over pollution in the environment from irresponsible disposal of waste products to a point where serious public concern is now evident. This has become particularly acute where passive low level exposure to deleterious materials over long periods in the normal living and working environment is suspected. The ultimate effect this may have on health remains difficult to assess.

The reasons for this are numerous, but two areas provide an insight. Firstly, medical and toxicological research. Toxicological research provides early indication of the possible health effects of new materials, but the relationship of such work to true environmental conditions is extremely difficult to interpret. Epidemiological research must of necessity lag behind material developments which inevitably means that the population's health may be at risk in the intervening period. Furthermore, this research is usually based upon workplace experience where exposure levels are generally higher than in the general environment. It is also difficult to separate the effects of a suspected material from the influence of a range of other environmental factors.

The second area involves the practical assessment of the applications within buildings. Here the difficulty is in assessing the dose or exposure an individual might receive. Figure 1.1 lists the interrelated factors effecting dose or exposure, which illustrates the problem quite vividly.

Figure 1.1 **Factors influencing dose or exposure**

Factors	Comments
Form and condition of material	Is the material loose and friable – will it be a source of dust? Does it contain volatile elements – will it emit toxic fumes by 'off-gassing'? Is it combustible – again will it omit toxic fumes? Does it contain naturally radioactive elements?
Position within the building	Contact with the water supply? Contact with foodstuffs? Internal or external? Exposed or concealed? Is there any danger from physical contact?

Means of degradation	Abrasion	- normal weathering - normal wear and tear - DIY activities (sanding)
	Chemical action	- corrosion - drying - gas emission - DIY (burning off)
Ventilation	Air change rate	- residual properties of dwelling - normal rates achieved by opening windows etc.
Lifestyle	Periods of occupation The time factor governing the period of exposure	
Maintenance cycles	May introduce toxic chemicals or increase dust resulting from maintenance	

1.2 ENVIRONMENTAL ISSUES

The environmental impacts of building materials are also difficult to assess primarily due to the many different issues that have to be considered. Figure 1.2, whilst not an exhaustive list, notes the various factors that can impact on any assessment. To make matters worse there are often inadequate data available to make an accurate assessment and even if there were, the implications to the designer selecting materials for a building will be further complicated by the source of materials relative to the location of the completed building.

Figure 1.2 **Key factors effecting environmental assessment of building materials**

Process	Issues
Upstream – extraction and manufacturing	Energy involved in both extraction and manufacturing processes. Transportation from source to manufacturing plant Depletion of resources – how much reserves remain? The amount of despoliation caused by the material extraction process Quantity of waste generated at both extraction and manufacturing stages Quantity of pollutants generated during these processes Proportion of the product made from recycled material
Construction	Energy involved in the construction process The distance the material has to be transported to site How much waste is generated and how much is, or can be recycled? What pollutants are generated during the process?
Buildings in use maintenance	Durability of the material in specific application and effects on Life expectancy
Downstream -demolition disposal and recycling	Pollutants caused during demolition Pollution as a result of disposal The volume of waste to be disposed of The distance material has to be transported either to tip or recycling point What proportion is recyclable or reusable? Ease of disassembly

1.3 SCOPE AND CONSTRAINTS

From the points itemised in Figures 1.1 and 1.2, the magnitude of the task facing the authors may be appreciated. Clearly it is difficult for one individual to possess the interdisciplinary skills to develop these ideas to a logical conclusion, hence the necessity for a group of specialists covering the fields of building design and production, environment, health and safety and medicine. The authors hope that their attempts at co-ordination have met with some success.

Whilst a review of the health effects and environmental impacts of all materials available for the use of all building applications forms a desirable, but unrealistic objective, clearly some compromise had to be accepted to retain the scope of the study within reasonable resource constraints. The study, therefore, has been directed as in the first edition towards low-rise residential premises, the reasoning being that this forms a large percentage of all buildings both old and new and at the same time is a clearly identifiable area of building technology. Further, it has implications for all sectors of society since, to all intents and purposes, we all live in houses or flats with approximately seventy percent being owner-occupiers.

It has also been necessary to rely upon the current 'state of the art', i.e. to base the study on current knowledge. The authors have not commissioned any special research nor has any special testing been undertaken. So reliance has been placed upon general knowledge of the constituents of building materials and on information supplied by manufacturers regarding their individual formulations. In the first edition the authors encountered considerable difficulty over disclosure of information. Much has changed since then and the majority of suppliers have a more responsible attitude in declaring the basic materials used in their formulations, mainly due to the statutory requirement to do so. However it is still necessary to request this information, as the standard trade literature is by its very nature produced to stress the best performance characteristics of the product. The authors have found a general willingness from manufacturers to co-operate when advised of the nature of the study for which we thank them, but as suggested in the first edition, if an abbreviated version of this type of information were included in the standard literature it would be of great help.

Since from time to time manufacturers alter the composition of their products from that studied by the authors. So it is suggested that in specifying any product where the health or environmental issue is of importance, the manufacturer or supplier be contacted if there is any doubt concerning the contents of the product.

1.4 FACTORS EFFECTING SELECTION

In considering the overall objective, it became obvious that a number of questions would need to be resolved when selecting a material or component for a specific application:

1. What are the main generic materials used in the building?
2. Where is the application in the building?
3. What alternative materials are available for the application?
4. Will the technical performance and appearance of the alternatives be adequate?
5. What are the comparative health hazards for all the alternatives?
6. What is the environmental impact of using each alternative?
7. What are the comparative costs?
8. What action should be taken when deleterious material is discovered in an existing building?

It should be noted that in attempting to provide answers to the above questions, the authors have restricted themselves to the materials that form part of the building fabric, services and fittings. Furnishings and loose furniture would require a further study. The one exception to this being carpets as these are often provided as part of the completed building.

Whilst questions 1 and 2 above are relatively easy to resolve the remaining ones are the most pertinent for the designer. Final selection will require them to make health, cost, environmental and technical comparisons between the alternatives. Questions 4, 6 and 7 also relate to life cycle analysis and costs which is another consideration the designer increasing needs to take account. This issue is dealt with in more detail in chapter 3.

1.5 ASSESSMENT – HEALTH AND ENVIRONMENT

If the building designer is to select appropriate materials to reduce hazards to health and the impact on the environment, in order to answer the questions posed in 1.4, it is necessary to make a sensible comparative assessment of the alternative materials available for the application in question. Thus assessment forms the crux issue and a great deal of the authors' time was devoted when writing the first edition to establishing a workable system, because without this, the study could not have proceeded satisfactorily. The debate at the time revolved around whether or not an absolute categorisation ('safe' or 'unsafe') or a relative scale of hazard would be most appropriate.

An absolute scale would have been ideal, but it became obvious that this would be impractical for two main reasons. Firstly the lack of complete and conclusive toxicological and medical evidence on many of the materials was such that subjective judgements, or at the very least extrapolation from applications from other industries or from occupational requirements, would be necessary. Secondly the matter was further confused by the need to take into account the position in the building and relate this to other factors such as the rate of ventilation, in order to estimate the possible 'dose' or exposure to the hazardous material. The problems of achieving this with any degree of confidence have already been identified in Figure 1.1. It would have been extremely difficult to establish the risk on an absolute basis, without undertaking exhaustive tests and trials. Little has changed from this perspective since the writing of the first edition.

When compiling this edition, a similar debate ensued over assessing environmental impact. As the work continued, it became clear that there were many gaps in the scientific knowledge needed to support all the decisions. The authors accept this deficiency and the impact it may have had on their conclusions. Therefore, as with assessing health hazards, an absolute scale was impractical and a relative scale unavoidable. Since the system for health hazard assessment proposed by Dr M Greenberg in the first edition, had proved popular due to its simplicity, the authors developed the system to take into account environmental impact. As before, it is clear that a relative scale is unavoidable and is adequate for comparative decision making at the early stages of design, so the same system has been adopted for this edition.

A hazard scale of 0 – 3, identified as:

> 0 – none reasonably foreseen
> 1 – slight/not yet qualified by research
> 2 – moderate
> 3 – unacceptable

In this edition, in the context of health hazard assessment, the scale is applied to two different categories as the long-term environmental impact previously considered in this assessment is now in the new section environmental impact assessment. These are defined as:

(A) The potential health hazard to the occupant when the material is in position in the building.

(B) The potential health hazard to the occupant when a reasonably foreseeable disturbance of the material could occur due to maintenance, repair, replacement or fire.

So, for example, asbestos cement slate on the A/B scale would rate 1/3. Category 1 because when it is fixed on a pitched roof of a building, asbestos fibre release will be of a low order in early years, and category three, because the risk of release on cleaning, maintenance or fire is unacceptable.

Whilst this provides a workable system, it should not be forgotten that very few normal activities are totally devoid of risk to health. Allergic reactions in susceptible individuals, such as hay fever and asthma, elicited by pollen, spores, house mites or certain types of food are well known. Perhaps less well known are those risks, which are associated with more serious diseases such as cancer. Such everyday processes as frying and grilling food produce traces of materials which in sufficient amounts have been shown to cause cancer, though the risks in the stated example are very small. Similarly some materials used in buildings have been shown to cause cancer under certain conditions. In assessing the safety in use of these materials, the authors have had to make a judgement whether the risk from the intended use was significantly greater than that encountered in normal everyday use. If in their judgement it was not, then it scored 'none reasonably foreseen'. However, because many people would prefer to avoid this risk, even, though it is very small, where it occurs, its existence has been recorded.

The problem of the type and nature of the health information has been already mentioned. As there are insufficient toxicological and medical data available on many of the materials currently in regular use and in particular the substitutes for materials this often means that clear-cut advice cannot be given. So the authors have taken a possible optimistic view using factor 1 on the risk assessment scale for materials where the risk is still not determined. However, to penalise a material which may prove perfectly safe causes an equal dilemma.

The authors accept that the use of these somewhat subjective judgements in coming to conclusions is not completely satisfactory.

If the risk of fire involving a material in a given situation has been significantly modified by the choice of an alternative material, then this itself will obviously have an influence on the overall risk to health of the occupant associated with the choice of material, particularly if there is some significant health hazard associated with the degradation of the material by fire. This has been taken account of in the risk assessment when deciding the hazard rating for category B of the risk assessment.

Similarly ventilation has been considered when appropriate and, like fire considerations, is not simply a problem of material selection, but is also influenced by the general building design and layout. Equally changing life styles can effect both relative humidity levels and ventilation rates – see 1.5 below. However very low ventilation rates (below 0.5ach) can be, in themselves, a health hazard so it is important to appreciate the interaction between a possible pollutant and the normal ventilation expected in buildings.

In the case of environment hazards the same 0 – 3 scale is applied to four main categories:

(A) The environmental impact upstream, that is to say the extraction and manufacturing processes.

(B) The environmental impact during the construction process.

(C) The environmental impact during the life of the building

(D) The environmental impact downstream, that is to say on demolition either as waste or in recycling.

In assessing the impacts, the authors applied the hazard scale to secondary issues as identified in Figure 1.2. These assessments were based on published environmental data on materials from research carried out elsewhere or upon the authors' own experience. Assessments were made for all the alternative materials under categories A, B, C and D whereupon a judgement was made as to the point on the hazard scale. Again, it is accepted that this is not entirely satisfactory. A more scientifically based life cycle assessment approach is possible but requires much more extensive research and for various reasons, explained in Chapter 3, is problematical in itself. It is important to note that the environmental impact grading is a **_relative_** assessment between the alternative materials identified for a particular application and not between applications. Therefore cross comparisons of gradings between data sheets should not be made.

The transport of materials from place of extraction to the manufacturing and construction processes, or direct to the construction site, gave the authors problems in fully considering the environmental impact. They concluded that it was impractical for them to give any definite answer and the user would have to consider and judge transport implications themselves when selecting a material where it was appropriate so to do. For example, there are well known cases where the extracted material was transported long distances to the place of manufacture only to be hauled back to the construction site close to the place of extraction.

1.6 EXAMPLES OF HOW TECHNOLOGY AND LIFE STYLE CHANGES HAVE INFLUENCED THE INTERNAL ENVIRONMENT OF BUILDINGS.

In the early seventies, the oil producing states of the Middle East commenced an oil embargo. One of the effects of this action was a wider awareness that a considerable amount of energy was being wasted in the operation of buildings. Resulting from this several actions were taken, including improving insulation performance, reducing the number of air changes and increasing the efficiency of heating equipment as well as air-tightness of the building. The later was achieved by a variety of measures including designing new houses without chimney flues, sealing flues in existing buildings, improving the air-tightness of windows and doors either in their design or by using draught exclusion products, and the development of the double and secondary glazing markets.

At approximately the same time there was a revolution in the availability of new building products, notably with the development of the petrochemical industries. Plastic products, new adhesives, solvents, fungicides and other chemicals were being used increasingly in the construction or furnishings of buildings. The majority of these, if not all, had had little or no scientific analysis carried out to determine the effects of their use on the occupants of the building or indeed any problems associated with the combination of such materials.

Resulting from these situations were room spaces within dwellings with much less ventilation, but including an increase in the number of pollutants.

One also must consider the changes in lifestyle that have occurred over the last fifty years. In the immediate post war period, the majority of householders lived without central heating and with either a polished or varnished timber floor or with linoleum with an occasional rug. As society became more affluent, linoleum gave way to carpet and by the late sixties, cheap foam-backed broadloom carpets were readily available. At this time central heating was becoming the norm in new housing and retrofit was also occurring on a substantial scale. During the seventies and onwards, the double glazing market expanded dramatically to such an extent that it is now the norm to install it in all new buildings.

The technology and life style changes effected the relative humidity in buildings. Prior to central heating, humidity levels would generally be higher. The advent of central heating initially reduced these levels which subsequently increased again with the installation of double-glazing and other airtightness measures. Add to this the reduction in air change rates mentioned previously and the question should be posed whether or not there is any correlation between the increased incidence of ill-health in terms of allergy and asthma. The house mite infestation implicated in the epidemic of asthma in the UK may be a symptom of changes in life style and technology which have occurred over the last half century.

The long-term effect of reduced ventilation rates and such other changes on the health of the occupants is still not fully understood. A number of other environmental and building technological changes and their implications are explored in Chapters 3 and 4.

1.7 TECHNICAL ASSESSMENT

For each application the materials have been compared against a performance specification which may exceed the minimum laid down by the Building Regulations, but is generally related to normal building design and practice.

The assessment used for the technical and aesthetic performance is a relative scale of 1 – 10. A simpler scale is adopted, as the problems associated with investigating the comparative performance are less severe, as designers regularly carry out this type of technical comparison. This scale is provided to assist, but it is anticipated that it does not cover every conceivable use of the materials. Designers must, therefore, use their own judgement in individual circumstances.

On the scale of 1 – 10, 1 represents the best material available, i.e. that considered from a technical point of view, and 10 to be generally unsuitable for the application under consideration or having poor life expectancy. In all cases the grading is a compromise considering all function and performance factors, durability and buildability, i.e. the ease of construction. It could be argued that the latter factor is of lesser importance, but this does effect the contractor's perception of the best material to use.

1.8 AESTHETIC ASSESSMENT

On the whole aesthetic judgements have not been an issue in the assessment made in the application sheets (Chapter 6). This parameter is, quite rightly, left to the individual designer's preference. However, where in the judgement of the authors an alternative presents a significantly inferior appearance, comment is made in the guidance notes.

1.9 BUILDING COSTS

The cost ranking figures provide a comparative guide for each separate application. The material having the lowest unit cost is allocated a base rank of 100 and all other material costs are compared with this. This is only used as an indicator as it is not always possible to compare like with like. For example, some materials are produced in fixed modular sizes whereas others provide greater flexibility.

1.10 USE OF THE GUIDE

In an effort to provide precise and useful information for those involved in the design process, the detailed results of the technical, health, environment and cost comparisons are shown in a concise manner in the application sheets in Chapter 6. Each sheet also provides overall guidance on the selection for new buildings as well as comments on the possible problems encountered with each application in existing buildings. Detailed guidance upon the use of the application sheets in provided in Chapter 5.

However designers and specifiers should resist the temptation to solely rely upon Chapter 6. In order to appreciate the detailed health risk assessments and environmental impact shown on the application sheets, it is necessary to have a broad understanding of Chapters 2 and 3, which consist of a review of the recognised health and environmental hazards posed by building materials, respectively. The importance of this general understanding cannot be overstated. Although architects and other designers may feel that this is another imposition on their already short design time, and perhaps even on their design freedom, it has become obvious that, in particular, the health and environmental issues must be given more consideration as subsequent remedial or removal measures resulting from unwise selection may ultimately result in claims for negligence.

CHAPTER TWO

Health Issues of Materials Used in Residential Buildings

M GREENBERG MB FRCP FFOM

2.1 INTRODUCTION

Materials used in the construction of residential buildings can subsequently present health hazards to inactive occupants or to active DIY enthusiasts, as well as the workmen involved in service, maintenance, refurbishment and demolition.

Even when the most benign and 'greenest', of construction materials have been specified and employed, the siting of the building, its original design or its subsequent conversion, can lead to imperfect indoor environmental conditions due to: poor acoustics; vibration; dysfunctional lighting; inadequate air exchange; extremes of relative humidity; and soil mediated toxicity. These conditions will impair the durability of the building and the health and comfort of the occupants. Important though these design linked conditions are, on the grounds of practicability the scope of this chapter is largely limited to the potential health hazards of certain construction materials to occupants.

2.2 THE STAGES AT WHICH CONSTRUCTION MATERIALS PRESENT HAZARDS

A comprehensive audit of the human and environmental health hazards of construction materials, requires to take into account their whole life cycle, including:

1. the human and environmental health impact of the processes involved in obtaining the raw materials (e.g. mining, quarrying, manufacture, synthesis), and in the disposal of mine and quarry spoil or unwanted by-products of manufacture or synthesis;
2. specific hazards involved in the transport of raw materials;
3. health, safety and energy aspects of the manufacture of the building product, which will concern not only the production worker but also members of the general public and the environment;
4. exposure of construction workers and local residents and release into the general environment, arising from the way in which it is reasonably foreseeable that, labelling and instruction notwithstanding, the product will in practice be inappropriately handled, cut to size or applied on site;
5. the hazards to occupants of commissioned buildings of the installed products
6. risks to professional and amateur plumbers, electricians, decorators and residents arising from, maintenance, servicing and refurbishment;
7. risks to tradesmen and residents, and the general environment, arising from the disposal of discarded materials during replacement or demolition.

2.3 THE PROBLEM OF IDENTIFYING HAZARDS FROM CONSTRUCTION MATERIALS INSTALLED IN HOUSING

The stock of houses in continuous occupation in the UK, constitutes a thousand year history of construction techniques and materials. To employ an anatomical analogy, the material of their skeletons will vary from daub, stone, rubble and timber of earlier constructions, to brick, concrete, steel, and breeze block. Their surface 'skins', and the connective tissues filling in the hiatuses, will largely be composed of the materials in vogue at the time of construction, but subsequent modifications to improve the safety and amenity of older buildings, will have introduced modern materials, that inter alia update their health hazards.

For example, by the end of the 20th century, in the interests of economy and safety, thousands of tonnes of asbestos based construction materials will have been installed in the UK, in new and old buildings.

Similarly, on the grounds of improved amenity and of public health, early plumbing innovations will have provided a clean water supply, that in exchange for the ravages of cholera and typhoid, substituted the insidious effects of excess lead uptake, until plumbers turned progressively to less toxic materials such as iron, copper and polymers. As a result of piecemeal modification of older systems, concealed runs of lead piping are still to be found.

2.4 TOWARDS A 'SAFE' INDOOR ENVIRONMENT

Until relatively recently, the major public health concern in the city was the airborne pollutants in the outdoor environment. This consisted of a mixture of, suspended particles, sulphur dioxide and sulphates, together with the oxides of nitrogen, derived largely from the burning of fossil fuels by manufacturing industry and transport. To these were added the combustion products from domestic fuels used for heating and cooking, and the seasonal surges in general environmental ozone, pollens and spores.

Disruption of travel by smog, disfigurement and destruction of masonry, soiling of clothes and households, and such overt adverse health effects as acute irritation of eyes, nose and chest, were accepted in large conurbations for hundreds of years, until the cost in terms of human lives was determined. This led to the introduction of 'Clean Air' legislation which produced a dramatic reduction of environmental pollutants, and drew attention to the significance of the pollutants contributed to the indoor environment by, cooking, heating, pets, house mites and agents derived from construction materials.

There are a few products, which under reasonably foreseeable use and misuse are utterly safe. For a number of materials, first awareness of their hazard came from the experience of the work force involved in mining and manufacture.

Environmental Inspectors have used occupational exposure standards to derive what they believed to have been safe indoor exposure levels for members of the public. The first misunderstanding was that exposure at and below the occupational standards was safe. An occupational standard might involve the pragmatic acceptance of what can be achieved by the best practicable means. By dividing the occupational standard by 4, they hoped to adjust pro rata for the difference between the 40 hour working week and the 168 week of the totally housebound. The prudent Inspector, would by rule of thumb rather than from scientific insight, divide again by 10 to allow for the greater sensitivity of infants and others largely housebound.

While the original occupational standard might consider an excess morbidity or mortality rate of less than 2 per cent acceptable, the consensus in the non-occupational setting came to require excess mortality to be less than $1:10^5$. On this basis, division of the occupational standard by 40, was a far from conservative measure.

For most compounds with acknowledged toxic hazards, not only are dose and response data severely limited, but their relationships remain to be accurately determined. For many compounds that are suspect carcinogens, in the absence of means for ascertaining with any degree of confidence exposures at which excess incidence of mortality levels are $< 1:10^5$, value judgements require to be made to determine what is an acceptable exposure standard. So far, few guidelines have been produced for environmental agents (Maynard et al 1997) (WHO 1987) (Raw et al 1995).

2.5 HEALTH ASPECTS OF SOME MINERAL BASED CONSTRUCTION MATERIALS

The nomenclature for mineral products used in construction is commonly imprecise so that care is required in their selection. In the design of buildings, the use of hazardous materials or the selection of safer alternatives, will be determined by an informed specifier. Subsequently, the methods of working with hazardous materials and the selection of replacement products will be determined by the state of knowledge of builders and DIY occupants.

Stone

Broadly speaking, the health hazard of a stone used in construction relates to its crystalline silica content and the amount of dust generated from it that is inhaled. Substantial exposure leads to silicosis, in which the lungs lose their normal elasticity causing the act of breathing to become laboured: the circulation of blood through the lungs is impaired, so that waste carbon dioxide builds up in the tissues, and sufficient oxygen fails to get to the tissues. Ultimately, the heart gets overburdened and fails.

In recent years, the excessive inhalation of stone dusts has also become causally associated with an increase of lung cancer (IARC 1997). Stones vary widely in their crystalline silica content and disease potential, with sandstone and true granite being particularly notorious for masons. While sandstones are usually highly siliceous, stone designated as 'granite' can vary widely in quartz content from a few percent to over 50 percent.

The hazard of these materials lies largely in quarrying and masonry work, but cleaning can also be risky if jetted abrasive dusts are used to remove grime. Dry sand blasting proved lethal for operatives, and wet methods were introduced, and in many, but not all industrialised countries, a variety of powders have been substituted (e.g. treated sand, haematite, garnet, coal slag), though experimental studies of these substitutes shows them not to be biologically inert.

Slate

Slate is rich in crystalline silica and as a consequence slate splitters, engravers and workers moulding slate dust are at risk of silicosis. Slate in the massive forms of tiles and slabs, presents no hazards to occupants of buildings or to DIY enthusiasts, either for silicosis or for cancer.

Clay products

Clay is fired and may be glazed subsequently for decorative purposes or to counter porosity, in the production of roof, wall and floor tiles, sanitary piping and a variety of structural and decorative bricks. The firing process transmutes a largely silicate clay with a relatively low intrinsic health hazard, into forms of crystalline silica (cristobalite and tridymite) that are especially hazardous. Hazards are restricted to the manufacturing processes, and do not affect construction workers or the occupants so buildings.

Concrete

Patching concrete will expose the operatives or DIY enthusiast to the irritating and sensitising effects of the cements and the additives used. The form of dermatitis attributed to a repeated skin contact with chromium ions present in cement presents a particular problem. Withdrawal from contact with cement will allow the skin to recover sooner or later, but unsuspected rechallenge with chrome ions in another context can lead to disabling relapse. Repeated exposures to other additives may also lead to skin sensitisation.

Certain concrete aggregates may contain a substantial amount of crystalline silica, which when the concrete is drilled, cut or scored, will expose the operative, and people down wind, to clouds of potentially toxic dust.

Vermiculite

This dusty particulate product came to be used in block form, and as non-combustible loose-fill thermal insulation between the joists in the domestic loft. Concern developed for the carcinogenic potential when naturally occurring asbestos and vermiculite fibres were detected in a number of specimens. This was reinforced by the adverse health findings in a study of a group of miners.

The nuisance value of the dust tracking through hiatuses in the ceiling into the rooms is bad enough without the long-term cancer threat. Fibre free material can be specified if the material is required.

Calcium Silicate

The term 'calcium silicate' covers a range of chemical compounds in a variety of physical forms, of natural and synthetic origins. Some are discrete particulates, while some have 'fibrous' forms. Fibrous 'calcium silicate' has been tested for carcinogenicity and found to be positive.
As with vermiculite, non-fibrous products may be specified. Older 'calcium silicate' boards were reinforced with asbestos fibre.

Gypsum
This material need present no health hazard. However, at one stage, plasters and plasterboard from certain sources were estimated to be capable of contributing an additional 10 – 50 per cent to the total dose of ionising radiation (radon) originating from the soil in certain districts, that was received by occupants.

Asbestos
The general term *'asbestos'* is applied to the fibrous forms of serpentine and amphibole minerals. While the amphiboles Amosite and Crocidolite went out of use or were widely banned a number of years ago. Millions of tonnes of the serpentine mineral chrysotile have continued to be mined annually.

Despite the reduction in the burden of asbestosis, because of its carcinogenic properties, the continued use of asbestos came to be increasingly questioned. The latest international consensus statement (1998) on chrysotile asbestos (WHO 1998) concluded unequivocally:

'Exposure to chrysotile asbestos poses increased risks of asbestosis, lung cancer and mesothelioma in a dose dependent manner. No threshold has been identified for carcinogenic risks.'

The working party also recommended that where safer materials are available for chrysolite, they should be considered for use, and was particular concerned about its use in construction materials.

While in the Developed World, legislation, litigation and prudence have led increasingly to asbestos being omitted in new construction, many older buildings contain large quantities.

Even where friable asbestos products are not exposed and are not subjected to attrition, the natural movement of a building and its air currents will lead to dispersion of fibres. Such tradesmen as electricians, telephone engineers, decorators, heating engineers and plumbers, and enterprising occupants engaged on DIY, are at risk from exposure, and they will release fibre to contaminate the household.

Encapsulation postpones the expensive removal process, and may be acceptable in certain circumstances as an economical and effective temporising option for the management of asbestos, deferring the final reckoning to when demolition takes place.

The concept of asbestos fibres being firmly locked into a cement or polymer matrix is advanced to support the case for the safe use of chrysolite. However, even if during the active lifetime of a building attrition is avoided, the matrix will ultimately undergo degradation and release its fibre content.

In 1999 the European Union issued a directive virtually banning the use of chrysotile asbestos, and shortly after the UK implemented it banning the import and use of asbestos from November 1999 with a few time-limited derogations. The new use of asbestos cement boards, panel tiles and other products was forbidden, as was the sale of second-hand asbestos cement products.

Non-asbestos mineral fibres
In a guidance note to Local Authorities (WHO 1998(2)), WHO considered that there were numerous products where fibres were irreplaceable, and discussed the question of substitution for asbestos in construction materials by such products as: glass, rock and slag wool; special purpose glass fibre; continuous filament; ceramic and refractory fibres.

The introduction of 'cotton silicate' (slag wool) into the Royal Naval Dockyards in the 1890's, led promptly to complaints of severe irritation of eyes, nose, throat and skin. Protective clothing was developed, but crocidolite asbestos was substituted soon after as being less troublesome. Some 60 years later when the asbestos hazard could no longer be overlooked, mineral fibre was reinstated. Within 10 years of this, experimental studies showed that man-made fibres shared many of the biological effects of asbestos, raising the spectre of carcinogenesis, reinforced when subsequent epidemiological study reported increases in lung cancer in workers manufacturing rock wool/slag wool. Consensus opinion (WHO 1998) was that on the basis of the available data, it was not possible to estimate quantitatively the risks associated with exposure of the general population to man-made fibre in the environment.

Nevertheless, as general and indoor fibre levels were deemed to be several orders lower than those associated with occupational disease, it was concluded that the possible risk to the general public was low if at all. They were however careful to add the caveat that, '...[MMMF] *should not be a cause for any concern* **if current low exposures continue**'. Study of the acute pollution levels in various parts of a household, and of their decay subsequent to blowing glass fibre into a domestic loft, has been conducted, but otherwise indoor monitoring of mineral fibre, has rarely been conducted under the reasonably foreseeable conditions of use and misuse, and in view of the potential cancer hazard involved it would be difficult to perform such realistic studies on ethical grounds.

The hazard of a mineral fibre relates in part to its dimensions. The probability of a fibre being inhaled and penetrating the airways to reach the lung proper (parenchyma) where the business of gas exchange takes place, or alternatively depositing on the surface of the airway (bronchus) where cancer occurs, largely relates to its aerodynamic property, which is determined principally by its equivalent diameter. In the experimental situation, once the fibre reaches the critical organ, tissue or cell, the probability of tumorogenesis relates to the actual length.

With drawn fibre, the diameter is more readily controlled to narrow limits, so that fibres with low probability of inhalation and penetrance and of low tumorogenic potential can be selectively manufactured. With fibre or wool produced by blowing and spinning molten mineral, fibre dimensions are less easy to control to cluster closely round the nominal mean diameter.

Despite experimental evidence of the speed with which a fibre acts at cellular level, with certain effects observed after only a few minutes' contact, there is a belief that the survival of a fibre in tissue (biopersistence) determines its tumorogenicity. If this were so, then like other fibre factors, it is likely not to be a quantal effect, and low persistence fibres could not be discounted as innocuous.

A few non-asbestos mineral fibres, including ceramic fibres have been identified as potentially hazardous.

2.6 FIBRES DERIVED FROM POLYMERS

A number of such materials have been introduced as replacement for asbestos and other mineral fibres, including: polyamide; polyolefin; carbon/graphite (derived from controlled polymer combustion); polyester; polyacrylonitrile; polyurethane; polytetrafluroroethylene; cellulose, polyvinyl alcohol (PVA). They may be found free or incorporated in cement or other matrixes. Their hazards have been considered in comparison with asbestos. (WHO 1993) (Department of Health 1998)

Cellulose is a natural polymer for which animal toxicity studies have been inadequate, and for which human epidemiological studies are wanting. The distribution of the fibre dimensions and hence the proportions that are inhalable and respirable, will vary with applications. It has been used in loose fill form for loft insulation, though combustibility, infestation, dune drifting and ingress through hiatuses into the living areas present problems. Even though printing ink has been removed from recycled paper, the addition of fire suppressant and preservatives will require to be taken into consideration in evaluating the health hazard.

Polyvinyl alcohol fibre dimension distributions have been described as likely to contain few in the respirable range. Although toxicity and carcinogenicity studies have been too limited for assurance, certain reviewers (Department of Health 1998) have guessed that carcinogenicity will be lower than chrysotile and are reassured that exposure to respirable PVA will be very much below 0.5 fibres/ml. Considerations of the significance of the inhalable fraction, with the potential airway effect of the fall out fraction, and the significance of the respirable fraction of a compound whose monomer is reminiscent of vinyl chloride and vinyl benzene, would recommend caution.

Para – Aramid fibres are in general too large to be of concern as inhalational hazards. Their unusual property of fibrillation has made them the subject of experimental study. After intraperitoneal injection in laboratory animals, mesothelial tumours were observed and after inhalation of substantial doses, some lung changes were found. An expert committee considered the carcinogenic risk to be less than that from chrysolite, noting exposure as likely to be less than 0.5 fibres/ml (Department of Health 1998).

Other polymer fibres are subject to the same consideration as the above materials. Their chemical composition, insofar as they relate to known carcinogens will raise suspicion, but the quantities and dimensions of such fibres as become airborne, either from loose or friable material, or by attrition from a matrix, will determine the probabilities, a) of their entering the airways and the lungs and b) the probabilities of tumour generation.

2.7 METALS USED IN CONSTRUCTION

Lead

Historically, the criteria of lead toxicity were such effects as convulsions, paralysis, anaemia, severe colic and an associated malaise. Latterly, concern has been for whether covert effects might occur at levels of intake below those causing frank disease. The initial response to a study reporting the impairment of intelligence and development in children in relation to their uptake of lead, was an attack on the researcher. When the dust settled, the dose-related impairments of scores of tests of intellect and behaviour were acknowledged. As a confident threshold for these has not been demonstrated, there was consensus that where alternatives existed, lead should be replaced.

Water pipes, solders and paint have been the principal sources of environmental lead contamination in older buildings. For adults, the major source of lead will have been in drinking water, but young children will readily take to chewing paintwork, swallowing leaded window putty, and licking their hands clean of paint-flake enriched house dust.

Rubbing down old leaded paint during redecoration, whether with dry or wet abrasive paper, will contribute to the domestic environment and be a hazard for occupants, if the dust is not contained and the wet slurry is allowed to dry out.

In contemporary buildings, the use of copper (omitting lead solder for jointing) and plastic pipes for drinking water, will have eradicated one source of lead, and the reformulation of paint and linseed putty to exclude lead will have protected against some of the effects of the catholic tastes of children.

Leaded lights offer a limited opportunity for children to ingest the metal, and while the risk is small, sentiment is against the use of lead strip and if used it is recommended that it be made inaccessible sandwiched in double-glazing.

Inorganic lead was omitted from domestic paint, and organic lead was reduced in petrol as a result of a sustained campaign in which activists successfully lobbied industry and government. As a consequence of this campaign, at all ages there was a marked reduction in lead intake from new sources. Only in the older property is there a residual legacy of lead requiring to be dealt with safely during renovation and demolition.

Chromium

The metal, whether used in plating for decoration and protection, or in a corrosion resistant alloy, does not present a toxic hazard. Its salts have been used in cement additives, where they cause skin irritation on contact, and may lead to sensitisation with persistent dermatitis. They are also of concern as carcinogens. While the main problem in the context of concrete is in the initial construction of the building, subsequent alternations and refurbishment will expose workers and enterprising DIY occupants to hazard of a lesser degree.

Zinc

The use of galvanised iron, plain or plastic coated, presents no hazard to occupants, even if the coat is breached. As with cadmium, the inhalation of fresh fume from gas cutting of galvanised metal causes acute flu-like symptoms, though unlike cadmium, the long term expectations for zinc fume exposure are good.

Copper

Copper met with in construction as metal or alloyed in brass, presents no health problem for occupants. Heavy exposure to copper metal fume or to its salts, have only been met in industrial activities.

Cadmium
Cadmium is a toxic element with acute and chronic effects on lung, kidney and bone. The metal used as a protective coating for steel, does not constitute a hazard to occupants. Flame-cutting during redevelopment or demolition has led to attacks of 'metal-fume fever', some of which have been fatal.

Cadmium has been included in 'hard solder', which liberates toxic fume if overheated. This has been specified for the installation of certain domestic apparatus, which, while it was without danger to occupants, constituted a risk to engineers who were required repeatedly to disassemble soldered joints for servicing.

Cadmium salts are employed as pigments in paints and in plastics, but where there is a possibility that they will leach out and be ingested by infants, they require to be excluded from the formulation.

Iron, Steel, Stainless steel and Aluminium
Although these metals under certain circumstances may be associated with adverse effects, employed in construction materials it is not reasonably foreseen that they would offer a risk to occupants.

2.8 TIMBER, WOOD TREATMENTS AND WOOD PRODUCTS

Timber
Timber contains a large number of intrinsic biologically active and versatile compounds, to which are added a variety of chemicals to prevent staining or other microbiological and pest attack. As a consequence it is not surprising that wood working has been associated with skin, lung and heart diseases.

To complicate matters, commercial nomenclature is not a specific as botanical classification, particularly for some hardwoods, for example teak (*Tectona grandis*) may be substituted by iroka (*Chlorophora excelsa*). This is important where toxicity is peculiar to a particular plant. Thus, dust derived from the box wood of the notorious *Gonioma Kamassi* which acts systemically and effects the heart, differs from that derived from *Buxus macowani*.

While generally speaking untreated wood in bulk presents no hazard to health, the build up of volatile compounds in the indoor environment from odoriferous timber may prove irritant to the eyes, nose or throat of the sensitive occupant. This may continue until the volatile agent is spent or ventilation is enhanced.

The quantity and fineness of wood dust will vary according to the species of wood, its degree of seasoning and the speed and method of machining. Inhalable particles generated on cutting and sanding, may have toxic, immunological and carcinogenic properties according to the specific species of plant. Carpenters and joiners are not just at risk from the toxic hazards of exotic hardwoods; the machining of Canadian Red Cedar is a common cause of respiratory disease. Occupants of buildings in which these woods have been used would in the normal course of events not be at risk from wood dust, unless engaged in DIY sanding, or being in residence while this is being carried out. Wood dust exposure may lead to immediate complaints of irritation of eyes, nose, throat and chest, with paroxysmal coughing, breathlessness and wheezing. These complaints may be short lived or continue long after cessation of exposure. Even when complaints have ceased, objective tests of breathing may show a degree of impairment. (Wood dust hazards may be added to by materials used as protectives such as paint.)

The original finding of excess nasal cancers in furniture workers machining native hardwoods, has been confirmed on a number of occasions. Subsequent studies of factory workers exposed to dusts of a variety of hard and soft woods, have reported various other cancers.

Wood preservatives
Chemicals applied to standing trees and felled timber to prevent staining and spoilage by insects and micro-organisms add to the natural toxic cocktail present in timber. There are a number of preservatives, of which organotins, pentachlorophenol and benzene hexachloride are examples, about which there is controversy as to whether they themselves are responsible for the complaints made by occupants, or whether the sore eyes, headaches and chest symptoms are caused by the solvents. The matter has not been resolved, though what is unequivocal is that

these agents are toxic and that some of them may be carcinogenic. As a consequence, if damp proofing and ventilation do not obviate the need for treating underfloor timbers, they and their volatile additives should be contained or ventilated away to prevent build up in the indoor environment.

Surface treatments
Polyurethane is applied to wood surfaces in situ for protection. Unless previously sensitised to isocyanates, and then only if occupation of the treated area is too soon after sealing, the occupant is unlikely to be affected. However, the professional and the DIY enthusiast sanding, sealing or resealing an expanse of floor will be at risk of developing distressing cough, wheeze and breathlessness, which may continue to be disabling for some time. The risk is enhanced by the application of the sealer from spray cans, and may not be obviated by the use of a more viscous preparation.

Ply, block and compound
The sheets and particulates in these wood products are bound by a resin. Even when well-reacted and volatile monomers are not liberated in the fresh materials, it is in the nature of the resin to break down over time, with the rate of breakdown varying as the conditions of curing and ageing and the temperature of the building. As a consequence, in a number of cases where surfaces have not been rendered impermeable, measurable amounts of formaldehyde have been detected. Quite high levels may be achieved in a room where untreated particleboard has been used to surface walls, floors and ceiling.

Formaldehyde vapour can irritate and sensitise eyes, nose and throat, and it has been determined by the International Agency for Research on Cancer that there is sufficient evidence to categorise it as an animal carcinogen, though there is limited evidence of human carcinogenicity; as a consequence it has been considered as a possible human carcinogen.

Although it is still questioned whether prolonged exposure to the low levels met with domestically constitutes a health risk, it would be prudent to minimise the risk in the design of products and of buildings.

2.9 PLASTICS IN CONSTRUCTION

Construction materials made of 'plastics' are based on a single synthetic polymer or a mixture of polymers, to which a number of compounds are added to provide colour, plasticity, stability, fire retardation, and resistance to degradation by ultraviolet radiation.

Freshly polymerised materials have been detected to have emitted unreacted monomer and other volatiles for some time after manufacture, and to varying extents the product in situ has continued to release volatile components.

The additives include non-volatile compounds that are known to be toxic and some are suspected of being health hazards: these can leach out in water or in physiological fluids (saliva and gastric juice). Consequently, the formulation of a plastic component should bear in mind its foreseeable use (for example as a container for water), and equally foreseeable misuse (sucking and swallowing by infants).

Among the materials that have been added to the polymer mix as a filler has been asbestos. Fibre release has been reported from vinyl flooring as a result of attrition. The ban on the use of asbestos in new construction materials will restrict this problem to old materials and to their disposal. In the event of fire, although much is made of toxicity of the breakdown products of the synthetic polymers to exposed occupants, it is questionable whether in practice the hazard is significantly greater than for natural polymers such as wood.

Polyvinyl chloride
The monomer vinyl chloride has been accepted to be a human carcinogen at low levels since 1974. Originally, the monomer was found to be emitted from the materials in store and following installation, reaching unacceptable levels when ventilation was restricted. Subsequent technological development has reduced the order of this problem.

Unacceptable concentrations of phthalate plasticiser have been detected in air close to the surface of 'Vinyl' flooring material, corresponding to the nose level of infants for whom this agent may be a particular hazard. As previously noted, asbestos has been a filler incorporated into the PVC mix.

Urea formaldehyde
This polymer is capable of liberating detectable amounts of formaldehyde vapour that cause irritation to the eyes, nose, throat and chest accompanied by weeping, sneezing, coughing and breathlessness. In addition to its irritant and sensitising properties, this agent has been classified as a possible human carcinogen.

Formaldehyde vapour from incorrectly installed foam in cavity walls, has affected occupants symptomatically in ways that might be anticipated, but more general complaints have also been attributed to it in certain episodes.

When composite woods (e.g chipboard) are used extensively and ventilation is restricted, irritant levels of vapour may originate from the binding resin.

Polyurethane resin
Fully reacted polymer after painting or varnishing will not affect occupants, but subsequent DIY resurfacing and refurbishment will put the enthusiast, who it is reasonably foreseeable will not have the training in their safe use, at risk. Chest discomfort, and paroxysmal cough, wheeze and breathlessness may persist for a long time after they have been induced.

Epoxy resins and glues
When repeatedly in contact with the skin may induce a troublesome dermatitis. This is not a problem for the occupant unless involved in the DIY use of these products.

Natural and synthetic rubbers
These are used singly or mixed in various combinations to produce sheeting for roofing, flooring and underlay, or as sealants and gaskets. In position they present no hazard to the occupant. In the event of fire, their pyrolysis products are not significantly more lethal than those of wood. Stripping of old sheeting will only present a hazard when mineral fibre had been incorporated in substantial concentration and the process releases it.

2.10 ASPHALT AND BITUMEN

Asphalt and bitumen are used as waterproofing agents and as adhesives. While they contain fewer carcinogenic compounds then the older coal distillate products, skin contact should still be avoided by DIY occupants using bitumen based products to make good cracks in the waterproofing. Otherwise, occupants need have no concern, though the smell of volatile and aromatic components contaminating the indoor environment may be disturbing.

In the past, some bitumen impregnated roofing felts had asbestos added. While fibre release ought to be guarded against during their removal in the embrittled state, it is unlikely that the hazard will be recognised.

2.11 WOOD PRESERVATIVES

While virtually all materials used in construction, including concrete, timber and wood products, and synthetic polymers, are susceptible to attack by bacteria, fungi and yeasts, only wood products are generally treated with preservatives. These are applied to timber to protect it during transport, storage and after incorporation in construction. Correct design and maintenance of buildings, by not producing conditions conducive to fungal and algal growth, reduce the need for preservative. All wood preservatives are toxic to varying degrees; once treated timbers are incorporated into construction, their immediate risk to occupants depends on the extent to which their vapours enter the domestic environment. This will be determined by such factors as, the productís volatility, temperature conditions and the architectural design. Subsequently, sanding of exposed woodwork, and the disposal of wood by burning, will disperse the toxic agent.

Generally speaking, compared with solvent based preservatives, those based on boric acid and borates, though far from innocuous, in use are the least hazardous to occupants.

Preparations based on chrome and copper arsenite have a range of toxic properties including carcinogenicity. Solvent based zinc and copper naphthenates volatise, potentially to expose the occupants to their toxic fumes. Earlier preparations of pentachlorophenol contained toxic by products with a wide range of adverse effects including carcinogenicity. While these extremely hazardous contaminants have been removed from the product latterly, pentachlorophenol remains a toxic agent to whose fumes occupants should not be exposed.

As there are a number of factors that may affect the risks posed by the solvent based preservatives, it is not meaningful to attempt to rank them. If it is necessary to employ them, it is prudent to contain them.

2.12 CONCLUSION

Relatively few materials employed in construction have ever been adequately tested, either experimentally or for their human health hazards. Further, even where there is reason to take care, the commercial nomenclature of products may not be sufficiently specific to give a reliable warning. The labelling of hazardous materials is of severely limited value: even if it leads to safe installation, years later the product may no longer bear a warning.

The problems of persons juggling with the economic, technical and safety aspects of materials employed in new constructions, and for those faced with working with old products to be found in old buildings are formidable. With *new constructions*, some help may be given by product labelling and from information sheets, but these may deal in generalities ('use with adequate ventilation', 'avoid skin contact') and not offer the detailed practical guidance required for safe management.

With *old buildings*, the safety of workers, occupants and the environment depends on awareness by the architects, surveyors, builders or tradesmen concerned of the potential for hazards to be concealed on the premises.

An encyclopaedic knowledge of building materials ancient and modern is essential to architects and others if the protection of workmen, occupants and the environment is to be ensured. If there is an awareness and a will, shrewd networking will gain access to a wealth of information that is to be had on a pro bono basis.

2.13 REFERENCES

Committee on Carcinogenicity. *Statement on carcinogenic risks of three chrysotile substitutes.* Department of Health. 1998. [http://www.doh.gov.uk/chrys.htm]

World Health Organisation. International Agency for Research on Cancer. IARC Monographs on the Evaluation of Carcinogenic Risks to Humans. *Silica and some silicates, coal dust and para-aramid fibres.* Lyon, IARC 1997.

Maynard, R & Gee, D. (1997). *Air and Health.* Copenhagen, WHO Regional office for Europe & European Environment Agency.

Raw, G.J. & Hamilton, R.M. *Building Regulation and Health.* Building Research Report Watford, Construction Research Communications.

Air Quality Guidelines for Europe. (1987) WHO Regional office for Europe. WHO Regional Publications, European Series No 23.

International Programme on Chemical Safety. Environmental Health Criteria Document 77. *Man-made mineral fibres.* Geneva, WHO. 1988.

International Programme on Chemical Safety. Environmental Health Criteria Document 151. Selected *synthetic organic fibres.* Geneva, WHO, 1993.

International Programme on Chemical Safety. Environmental Health Criteria Document 203. *Chrysotile.* Geneva, WHO. 1998.

World Health Organisation. Asbestos and Health. WHO Regional Office for Europe, Copenhagen 1998(2).

CHAPTER THREE

Hazards to the General Environment From the Manufacture, use and Disposal of Buildings, Building Materials and Components

including the emerging requirements of the sustainable development agenda.

S.R. CURWELL, BSc MSc ARIBA
C.G.MARCH, BSc(Tech) MCIOB
S. NICHOLSON, BA. MSc

3.1 INTRODUCTION

In the introduction to this volume the overall environmental problem was subdivided into three general areas of concern:

1. **Global Issues:** Those affecting the whole planet and of importance to the global community.

2. **Local Issues:** Those of importance to individual nations or to the local community.

3. **Health Issues:** Those to do with the health and well being of individual users of buildings.

Chapter 2 has addressed the main health concerns related to building materials. This chapter serves to provide a very brief introduction to the global and local issues, particularly in terms of the impacts generated by the manufacture, use and disposal of buildings, building materials and components. It explores some of the primary selection factors and good practice guidance for reducing the environmental impacts of building materials and components. Reference is also made to specific methods that have been developed to provide structure and consistency in these decision-making processes. The relationship between sustainable development and the built environment is also explored in this chapter, as a brief introduction to this complex debate.

Figure 3.1: **Current environmental concerns**

Global issues:
- Global warming
- Acid rain
- Ozone depletion
- Deforestation
- Loss of bio-diversity
- Resource depletion

- Deforestation
- Access to 'green' space
- Noise (from the building process and from adjoining owners or users)
- Legionnaires disease
- Radon
- Electro-magnetic radiation

Local issues:
- Contaminated land
- Solid and liquid waste and landfill
- External air quality (nitrous oxide, sulphur dioxide, particulate matter, etc.)
- Water quality (drinking water and watercourses)
- Ecology and bio-diversity, flora and fauna
- Desertification

Health issues:
- Sick building syndrome
- Asbestos and other fibrous materials
- Indoor air quality
- Volatile organic compounds (VOCs)
- Drinking water quality
- Radon
- Electro-magnetic radiation

Designers, specifiers and builders are becoming increasingly sensitised to the broad range of environmental problems, as identified in Table 3.1, but are confronted by a bewildering array of possible actions and solutions. Each issue requires scientific and in some cases international political understanding to fully appreciate the complexity of cause, effects and proposed actions. Nevertheless it is important that construction professionals have sufficient general awareness and understanding in order to more fully appreciate the information and guidance given in the data sheets in Chapter 6. Therefore the overall aim of this chapter is to provide a foundation to the wide range of environmental and sustainable development issues.

3.2 GLOBAL ISSUES

There is mounting evidence that the global climate is changing, and that it is changing as a result of human activities. In particular, it is our increasing levels of emissions of gases such as carbon dioxide (CO_2), methane, chlorofluorocarbons (CFCs), hydrofluorocarbons (HFCFs), nitrous oxides (NOx) and sulphur dioxides (SOx) that are causing the various changes. These gases are responsible for a range of now commonly documented environmental impacts including global warming, the production of acid rain, and the depletion of the ozone layer.

Global warming refers to the warming of the lower atmosphere though the 'greenhouse effect' and has been the focus of much debate in recent history, with many predictions being made about long term impacts on both natural and human environments. Although these predictions have varied in terms of their magnitude, significance and time-scale, similarities in the type of impact being predicted are present. Common themes include the melting of the polar ice caps, rises in sea levels with consequential loss of low lying land areas, shifting of climatic zones and spreading desertification. Currently available data supports these general predictions such as an increase in global temperature of around 1.0-3.5 degrees, 15-95cm rise in sea level by the end of the next century, as well as increasing occurrence of dramatic weather patterns (DETR 2000-1). In order to counter these undesirable environmental effects, a number of international agreements aimed at reducing greenhouse gas release have been put in place. The main agreements are the United Nations Framework Convention on Climate Change (UNFCCC) agreed at the first Earth Summit held in Rio (UNCED 1992) and the Kyoto Protocol in 1997. The voluntary action agreed under UNFCCC encouraged developed countries to return their emissions to 1990 levels by 2000. At Kyoto legally binding targets were agreed for various nations which, taken together, require a total cut of 5.2% in emissions by the developed countries by 2012. The UK has agreed to a reduction of 12.5% as part of the overall requirement for the E.U. of 8% (DETR 2000-1). These targets are considerably lower than a number of environmental researchers and NGOs think is necessary in order to allow the developing world the capacity to industrialise their economies to even a limited extent without further increases in greenhouse gas concentrations and in order to avoid serious environmental consequences. In this context a reduction of 50% or more of current consumption by the developed nations is considered necessary. Nevertheless whatever targets are thought to be most appropriate, it is almost universally agreed that action should be taken if only on the basis of prudence, i.e., what has become to be known as the precautionary principle.

Targets like these can best be achieved through a reduction in our use of fossil fuels (coal, oil, gas) as their combustion is responsible for the release of large quantities of CO_2, a greenhouse gas, as well as a selection of other harmful gases. Different methods of fossil fuel combustion will result in different levels and types of emissions, depending on the specific fuel, combustion process, and exhaust and gas scrubbing systems employed, so some systems will claim to cause less pollution than others. However, taken overall, the most effective method of reducing CO_2 emissions is to reduce demand for, and consumption of, energy.

The use of the built environment has a key part to play in achieving these reductions as the majority of fossil fuel combustion is related to the consumption of energy for heating, lighting, servicing, etc, in buildings, and for the transportation of people between buildings. For example, buildings and the associated transport activity between buildings, in the world's affluent countries, is estimated to be responsible for around 60-70% of the total energy consumption. Thus, energy efficiency is a key requirement for reducing CO_2 emissions in the built environment. It is a very high priority for consideration in building design, location and use, the manufacturing processes used to produce building components and materials (their 'embodied energy') and the construction processes used to make buildings.

Acidification. Other harmful gases emitted through the combustion of fossil fuels for energy, particularly electricity, include sulphur and nitrogen oxides. These chemical compounds combine with water vapour in the atmosphere to form acidic aerosols, which are precipitated as acid rain.

This can cause environmental damage to forests, water courses and the soil, ultimately leading to the death of those animals and plants which cannot tolerate the increased acidity. Acid rain has also been known to cause damage to the built environment itself by increasing the erosion of porous stone buildings and the corrosion of ferrous metals. Again greater energy efficiency in order to reduce the emissions of combustion gases can be seen as a key requirement in the reduction of acid rain resulting from the production and use of the built environment.

Ozone Depletion. Another issue of global concern, that is often confused with global warming, is the so called "ozone hole", i.e., the damage that is being done to the stratospheric ozone layer that protects humankind, animals and plants from harmful ultraviolet (UV) radiation from the sun. There is clear evidence that the ozone layer in the upper atmosphere has become seriously depleted in recent years, about 7% reduction/decade in the Northern Hemisphere has been measured (SORG 1996). Some sections are becoming so thin at times that they are referred to as "holes", for example over Antarctica where the density in winter is 60% less than that observed in the 1960's (SORG 1996). This damage is being caused by emissions of man-made chemicals containing chlorine and bromine, such as Chloroflourcarbons (CFCs), Hydrochloroflourocarbons (HCFCs) used in refrigeration, air-conditioning (AC), foamed insulation, Halons used as fire suppressants, Carbon Tetrachloride and Trichloroethane used as solvents in paints, adhesives and sealants. International action has been taken since 1987 to reduce this damage through the "Montreal Protocol on Substances that Deplete the Ozone Layer" (UNEP 1987). In fact the original protocol has been amended and strengthened on a number of subsequent occasions as scientific understanding developed and the seriousness of the environmental problems became more apparent (London 1990, Copenhagen 1992, Vienna 1995). The production and consumption of CFCs has been effectively banned in most affluent countries and the development of a range of 'ozone friendly' chemicals suitable for substitution in products that once required CFCs has taken place. Unfortunately, some of these replacement chemicals, e.g. HCFC 134a, have been deemed to be potent greenhouse gases and their containment is recommended on these grounds. It is anticipated that new E.C. regulations will result in the progressive phase out of HCFCs by 2025, with use restrictions in the intervening period (OJ 1998).

The building industry must accept responsibility for a proportion of the damage that has been done as approximately 50% of all ozone harmful CFCs produced throughout the world were used in buildings (Vale and Vale, 1991). Although acceptable substitution products have been developed and are in use in many buildings, the expense of conversion means that a great number of AC systems continue to use the ozone damaging CFCs, even in countries where production has ceased. An extremely profitable black market has therefore been created, exemplified by the fact that the smuggling of CFCs into the USA is said to be second only in $ value to that of cocaine. So emissions continue to be a problem through illegal use and the fact that a significant number of nations (e.g. China and some of the former "Eastern Block" countries) are not signatories of the "Montreal Protocol". Thus, the use of certain foamed insulation products, the methods by which they are installed, and the provision of AC systems in buildings has been an issue of concern to well informed designers and engineers for some time in the light of ozone depleting impacts. Certainly, AC is rarely used in housing and flats in the UK and is not necessary for most buildings in temperate climatic zones as careful design can provide adequate ventilation, eliminating the need for mechanical cooling in the vast majority of spaces, and can result in greater energy efficiency. CFC-free insulation materials and methods of installation have been readily available for some time. Therefore there is a risk of complacency over this issue. It may be seen as a problem already solved. However it is important to continue to control CFCs, HCFCs and the other solvents. Even if all releases of these materials were effectively stopped in all nations, and this is far from the case at present, it will take around 100 years for the ozone layer to return to the level of the late 1980s (USEPA 1988)

Deforestation & Bio-diversity. Forest cover is another issue of global concern as both the temperate and tropical forests contribute to the regulation of our climate and fix CO_2 from the atmosphere. Loss of forest cover, or deforestation, contributes to global warming by reducing the capacity to fix CO_2 from the atmosphere and at the same time releasing the CO_2 embodied in the timber. This is due to the fact that a significant proportion is burned, as two fifths of the world's population depend on fuel wood for domestic energy needs (UNFAO 1997). The forest cover declined at the rate of 15.5 Mha/annum in the decade 1980-90, which slowed to 13.7 Mha/annum in the period up to 1995 (UNFAO 1997). The developing countries (mainly in the tropical forest regions) are currently losing 0.8% of their natural forest/annum. This is primarily through cut and burn to create extra space for farming and for fuel wood to support increased human activity. Demand for fuel wood is growing at 1.2%/annum (UNFAO 1997).

Forests are also exploited by the lumber industry who have traditionally demonstrated poor environmental practices. For example, despite the fact that only a small proportion of the trees in any particular area of tropical forest are suitable for use in building, for furniture or other industrial purposes, complete cutting of virgin forest is common, where trees unsuitable for lumber go for paper pulp or firewood. This type of practice not only affects the global climate but also threatens the natural habitats of many species, causing a loss of bio-diversity. Development of roads necessary to extract the timber is also a contributory factor in aiding human migration into the forest.

Practice in temperate forests is usually better, with many being managed through the use of good husbandry and replanting schemes. The North American temperate hardwood forest has recovered to a very considerable extent in the period since 1900. However, in some managed softwood forests it is commonplace to find monoculture practices that are unlike natural habitats and so have a relatively low ecological value; thus, they too can contribute to the loss of bio-diversity.

The issue of timber use, therefore, provides designers and materials specifiers with a dilemma. In principle, timber is a renewable resource, and therefore a very desirable, environmentally friendly building material. It is generally considered that the majority of softwood from European sources is sustainably managed, as is temperate hardwood from North America. Tropical timber is more problematical. A substantial proportion of current forestry practice is still purely extractive, i.e. full-cut without replanting or regeneration, and therefore contributes to both global warming and loss of natural habitats and bio-diversity. The International Timber Trade Organisation (ITTO) year 2000 objective, set in 1991, stated that "ITTO members will progress towards achieving sustainable management of tropical forests and trade in tropical timber from sustainable managed sources by year 2000" (ITTO 2000). It was anticipated that this would include some form of guarantee, through a labelling or certification scheme to assure buyers that timber was harvested from a well-managed, sustainable source. Unfortunately progress has been slow. In a report published in 2000 only six countries, Ghana, Guyana, Indonesia, Malaysia. Cameroon and Myanmar were considered to be managing some of their tropical forest reserves sustainably (ITTO 2000). Alternative certification schemes have developed (in the absence of one promoted by the ITTO) to label timber as having come from a sustainably managed source. The best of these is that promoted by the Forest Stewardship Council, which has received considerable international approval and recognition and forms the best available information at the present time (FSC 2000). The Council, through a number of sister organisations, now certificates a total of 240 forestry organisations in 33 countries, managing a total of 18,286,298ha. The wide differences in standards of practice in the lumber industry in terms of environmental performance mean that the source and type of timber must be considered as a key selection factor if deforestation and loss of bio-diversity are to be reduced.

Resource Depletion. Excessive fossil fuel consumption and the unmanaged use of timber are examples of the ill-considered use of resources that can cause environmental stress of global significance. Use of many other natural resources, such as minerals, can be just as damaging. Factors to be considered include whether a material is non-renewable, or finite in stock, and the extent to which it has already been depleted. This issue has assumed increasing importance in the quest for more sustainable development and will be returned to later. Mineral extraction and processing also cause important impacts locally.

3.3 LOCAL ISSUES

The manufacture, use and disposal of buildings, building materials and components contributes not only to the generation of global environmental impacts - 10% of UK energy consumption is used in the production and transport of construction materials (DETR 1998), but also to a variety of locally significant environmental impacts. The main area of concern at this level is the occurrence of pollution to land, water, or air which has, in the past, been caused by the extraction and manufacture of building products, and during the construction and demolition of whole buildings. Impacts also accrue through the transport of material, components and labour to local areas for building projects. Activities associated with extraction, including quarrying, dredging, mining, forestry, the construction of water reservoirs, etc., have all resulted in damage, sometimes irreversible, to the landscapes, natural habitats and flora and fauna of the local areas in which these processes have taken place. Extraction and manufacturing processes have also been known to generate large quantities of solid, liquid and gaseous waste that has resulted in local incidents such as contaminated land and polluted air and watercourses. The introduction and implementation, in most affluent countries, of various legislative and regulatory measures has served to prevent the majority of local, previously polluting practices from being undertaken in the extraction and manufacturing sectors. However, environmental impacts continue to be generated by the built environment beyond these early stages.

The transport of products and labour to site, the construction of buildings, their maintenance, use and disposal can also produce polluting wastes. Construction generates approximately 70 million tonnes of waste/annum in the UK (17% of the total UK waste arrisings) the majority of which goes to landfill (DETR 1998) and is the most frequent industrial polluter. A local or regional area may also feel the impact generated by a specific building project through the manner of use of land. Particular actions and choices made about the use of land for built environment, e.g. to use greenbelt for new housing, cause loss of natural habitats and bio-diversity as well as reducing or restricting access to 'green' space, at the local, micro level. A number of these points as well as other local issues identified in Table 3.1 are explored in Chapter 4 dealing with the performance of the whole building.

The best method of minimising the local pollution impacts is to avoid them. Doing more with considerably less material thereby improving the overall resource efficiency of building construction, building material and component manufacture, and minimising all forms of waste that are generated throughout the full life cycle form one of the major challenges to the construction industry in the next decade. Such action is essential not only to reduce the amount of environmental damage that is felt at a local level, but also as part of the industry's contribution to more sustainable development, explored further later in the chapter.

In the context of resource efficiency and pollution minimisation a factor that may be considered as a broad basis of selection of materials and or contractors, is the presence of an Environmental Management System (EMS) in a company whose construction products or services are being considered. An EMS is designed to help a company identify the environmental impacts that its activities generate, to control all the significant impacts and structure the continued improvement of its environmental performance. Certain EMSs have received widespread acceptance and can be accredited to a recognised standard. The most common of these are ISO14001 and EMAS. Thus, if a company can show it has an accredited EMS, its environmental performance should be in a process of continual improvement and its products may therefore cause less environmental damage. Through supply chain management companies can also have considerable beneficial influence in reducing the environmental impacts, by requiring better environmental performance standards from all their suppliers.

To summarise then, the most important selection factors to reducing local environmental impacts are:

■ The level and type of pollution resulting from the extraction, manufacture, use, and disposal/demolition processes, including solid, liquid and gaseous waste.

■ The overall material use efficiency of the manufacture, construction and use processes, including their recycled content and potential for reuse or recycling.

■ The overall energy efficiency of the extraction, manufacture, use and disposal processes, including any contribution from renewable energy sources.

Environmental factors like these have become increasingly recognised in recent years by the building industry in general and various aids have been developed to enable their incorporation into building design and material selection processes. The development of this type of aid to decision-making is, however, fraught with difficulty. The primary reason being that pollution and energy use can and do accrue at each stage of a products life cycle, making a comprehensive analysis of any particular material, component, assembly or whole building immensely complicated. Various tools have been developed and are in use throughout the world based on the principles of Life Cycle Analysis.

3.4 LIFE-CYCLE ANALYSIS [LCA]

LCA is an analytical tool that attempts to assess the material content and majority of environmental impacts of any manufactured item. In construction it can be used to assess materials, sub-components, components and whole buildings. To achieve the fullest understanding of all the impacts for a particular component in a building, application requires considerable research effort, particularly for complex components such as a prefabricated window assembly or an item of mechanical services equipment. Deep analysis on the basis of LCA is normally outside the time and resource constraints of routine building design and material selection and is, in most cases, beyond the normal skills and experience of most construction professionals. There have been calls for manufacturers to make LCA information on their products more generally available, but it is important to note that at present full LCA is not required of manufacturers, although it may be undertaken as part of an EMS process. Also, as there is no international agreement over LCA techniques, manufacturers have concerns over inter-comparability of assessment - the 'level playing field' concept. In the UK the BRE is making progress with its Environmental Profiling System with around 20 material and component areas analysed (BRE 2000 –1)

The various LCA systems that are in use and under development are being used to underpin new building and material analysis computer based tools such as the BRE's ENVEST (BRE 2000-2) in the UK or the BEES 2:0 programme in the USA (BEES 1998). Such tools make a significant advance over non-LCA based tools. Firstly, from the point of view of the user, they simplify a complex analysis and, secondly, they provide output that is specific to the project in hand rather than the generic information that otherwise is all that is available to aid design decision making. Designers need to use and evaluate these new systems. Despite these new developments a range of questions still arise in LCA, i.e.:

■ the level of detail that should be included, e.g. in a complex component should all the environmental impacts of all the sub-components (such as fixing screws) be considered?

■ the spatial and temporal domains over which all the pollution impacts should be considered, e.g. 60 years for the whole building, 30 years for clay roof tiles, 10 years for a central heating boiler, etc.; and

■ the relative weighting of the various impacts, e.g., how seriously a global impact, such as ozone depletion, is judged in relation to a local impact, such as river pollution, in coming to overall conclusions.

The last point is usually addressed by asking users to judge the importance to them, their client(s) and/or to the situation in hand and to express this in the weighting factors. In this way these tools assume awareness by the user of the broad range of environmental issues, which may not always be the case. The weighting selected may have considerable influence and, in some cases, a predominant influence on the outcomes (Curwell et al 1999).

These difficulties have influenced the approach adopted by the Authors of this book. The information, scoring and guidance contained in the data sheets included in chapter 6 of this text are not based on full LCA analysis. Rather, they are based on a simpler assessment based on a review of LCA and other data that is currently available in the public domain and on the judgement of the authors. This assessment focuses on four main phases of the life cycle of a building:

■ Upstream phase, extraction of raw materials, transport to the place of manufacture and in the manufacturing process.

■ Construction phase, the transport to and incorporation of the material or component into the building.

■ Use phase, the durability, level of maintenance and/or renewal or replacement.

■ Downstream phase, the demolition, dismantling, reuse, recycling and final disposal phase.

In each of these phases existing data on the three areas identified earlier - the level of solid, liquid and gaseous pollution and waste, overall material use efficiency and the overall energy efficiency - has been used in assessments. Judgements have also taken into consideration the volume of material used in an application, its expected useful life, and the feasibility of, and ease of, reuse or recycling. For example, a small impact generated at the production stage may become significant if a very large quantity of material is used, it has a relatively short life and it is difficult to recycle. Alternatively, what appears to be a significant impact generated at the upstream phase may actually be reduced if the quantity of material is very small, it is long lived and is recycled.

3.5 WHOLE BUILDING ENVIRONMENTAL ASSESSMENT

Various tools or methods (offering less than a full LCA of the building) have been developed over the last ten years to assess the overall environmental performance of a single building, e.g. BREEAM, (Baldwin et al, 1998), Green Building Tool (Cole and Larsson 1997). In many respects these are attempts to formalise environmental impact analysis of a building usually to ensure conformity of approach in order to grant a certificate or label. This is very necessary where claims of "greenness" need to be substantiated and compared. These type of tools usually attempt to assess environmental performance to a greater or lesser extent in terms of the issues in Figure 3.1. This is often achieved via indirect factors, e.g., the operational energy consumption of the building as an indicator of CO_2 or comfort factors in some of the rooms or spaces as an indicator of health of the occupants. In terms of building materials and components the early systems usually consisted of a simple checklist of materials to be avoided, e.g. CFCs in AC, or VOCs in paints. In some cases limited performance dimensions, such as embodied energy are included. These methods struggle to adequately address complex issues such as the contribution that a material or component might make in enhancing the performance of a whole building. For example, the use of thermal insulation, that may cause pollution or is energy intensive in production, that will nevertheless reduce energy consumption over the whole life of a building are still matters for the judgement of the specifier. This is an ongoing area of research as these tools continue to be improved through use and refinement. The new version of BREEAM overcomes some of these problems with a much more extensive assessment of materials used in elements of the building. The BRE ENVEST programme for Office Buildings illustrates what can be done where LCA data underpins a whole building performance design tool.

3.6 SUSTAINABLE DEVELOPMENT

Sustainable development addresses how humankind can meet its present needs, within the limits of natural resources and systems, whilst ensuring that future generations can also meet their needs within the same system. The concept of sustainable development has been used at an international level for at least 20 years with one of its earliest mentions being in the World Conservation Strategy of 1980. A number of reports and high-profile commissions in the 1980's such as the World Commission on Environment and Development (1987) developed the concept. Events like the Rio Summit in 1992 (UNCED, 1992) brought it to still wider audiences and greater international recognition.

The popularity and importance of the concept is due, of course, to the recognition that human development and its relationship to the ecosystem must be managed in a more holistic manner. This interest is based not only on economic concerns, but is also in part attributable to the fact that the meaning and concept of sustainable development has proved notoriously difficult to define. Many different definitions have been proffered in many contexts over recent years, giving various sectorial interests the option to adopt the one that most suits their own (partial?) view of the overall agenda. These differences in interpretation arise for two reasons. Sustainable development is inherently a value-laden concept as it is applied in the development activities of specific human societies and communities involving all stakeholders, and not in a social vacuum. It is therefore interpreted and given meaning in terms of the particular value system of a society. Secondly, sustainable development is new 'science' and because of its far-reaching implications

Figure 3.2 **The Principles Underlying Sustainable Development**

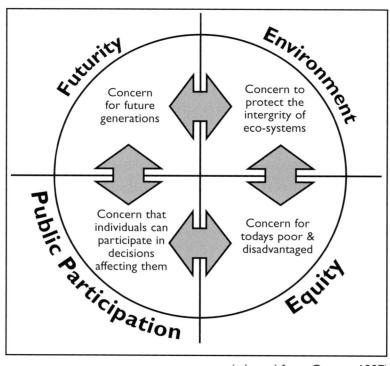

(adapted from Cooper, 1997)

requires much more interdisciplinary understanding and co-operative action. There is, thus, an intimate relationship between the principles, beliefs, ethics and values underlying any particular society or community, and the collective approach to sustainable development. The issue of a common language of sustainable urban development is the subject of a major E.U. funded project called BEQUEST. (BEQUEST 2000)

Despite these complex characteristics, the concept is still heralded as a coherent policy objective by national and local governments, retains its status as an internationally accepted goal, and continues to inspire much research and literature. In an attempt to better understand sustainable development, Mitchell et al (1995) reviewed some of the burgeoning literature on the subject and identified four common principles (Cooper 1997). These are listed below and shown diagrammatically in Figure 3.1, which also emphasises the interdependence of the four principles.

1. 'Futurity': i.e., the need for present generations to ensure that future generations have the ability to maintain current standards of living whether through the exploitation of natural or cultural capital. In other words, we must preserve natural and cultural resources, and we are only entitled to use up finite resources if, in so doing, we provide to future generations the know-how (e.g. through improved science, technology, social organisation, etc.) to maintain these living standards from the resources that are left. Otherwise, we are "cheating on our children".

2. 'Environment': i.e., the need to preserve the integrity of eco-systems, both at the local level and at the scale of the bio-sphere, in order not to disrupt or irreversibly damage natural processes essential to human life and the protection of bio-diversity.

3. 'Public participation': i.e., the need for widespread stakeholder participation in the process of change towards sustainable development, falling in line with Principle 10 of the Rio Declaration on Environment and Development (UNCED, 1992):

   "Environmental issues are best handled with the participation of all concerned citizens... each individual should have ... [information], and the opportunity to participate in decision-making processes."

4. 'Equity': i.e., the recognition of the principle of fair shares, to be implemented both locally and globally, as reflected, for example, in Friends of the Earth' s strategy for a sustainable Scotland (1995)

   "If there is a finite amount that we may consume or use beyond which we cannot go ... then we must share what we already have far more than is currently the case. Equality of access to the world's global resources therefore must be the guiding principle."

Work by the BEQUEST Network (BEQUEST 1999) has shown that the majority of construction professionals have an inadequate appreciation of the sustainability 'problem'. They tend to see it primarily in terms of the first and second principles, i.e., in terms of environmental protection. This is potentially very disabling as it indicates a relative lack of understanding of, or a lower interest in, the wider agenda of economic and social change that is very likely to be necessary in the affluent countries if sustainable patters of life are to become a reality.

3.7 THE WIDER AGENDA

Towns and cities are the containers of the vast majority of commercial and social activities, providing the reason for and the context of most human activity, of our work, recreation and leisure. The buildings together with the supporting infrastructure, including transport systems, utility services and communications form the base of modern civilisation and so dictate many aspects of human development influencing many lifestyle choices. Thus fuller appreciation that sustainable development has much wider social and economic dimensions beyond the 'technical' problem of environmental protection is the first step. The economic, environmental and social gains and losses of any particular re/development have to be weighed (and balanced?). Who is to gain and who is to lose out? Many questions arise, such as:

■ Will a new building assist local employment both during construction and afterwards?

■ Will construction materials and components be manufactured locally?

■ Will a new development aid inward investment or improve tourism (e.g. a new museum)?

■ Will resident's quality of life be improved or degraded?

■ Is the loss of an amenity (e.g. a green space) to be offset by creating a new one nearby?

■ Is the built cultural heritage to be enhanced or degraded?

Many of these points have been important to the control of development in the past, but take on much more significance in the context of sustainable development. This also raises the question of who should make the decisions. In the UK urban decision-making is mainly the province of developers, planners and local politicians. Is this adequate in the new situation?

In the UK Local Authorities were expected to develop a Local Agenda 21 (LA21) strategy by January 2000. Strategies usually identify a number of environmental, social and economic issues, which any LA (on behalf of and with the help of the local community) intends to address. Table 3.2 provides an indication of the range of issues often considered and illustrates the scope of the 'wider agenda' mentioned above.

Table 3.2 **Some local indicators commonly used in the U.K.**

Water consumption	Home composting
Energy consumption	Renewable energy
Production of waste	Travel patterns (e.g. less car use)
Air Quality (or pollutants)	Access to the natural environment
Water quality (or pollutants)	Accidents
Indicator species (flora & fauna)	Crime
Land use efficiency	Unemployment
Noise	Asthma
Recycling (amount or facilities)	

Adapted from ROTHEROE, C et al, 1997.

Factors such as crime or noise may at first seem irrelevant to sustainable development. However it is recognised that in some inner-city areas high crime rates or excessive noise are factors in reducing the overall habitability and represent some of the characteristics of **unsustainable** communities. Defined as:

■ a physical environment which has degraded and become polluted, with an overloaded or degenerating and inefficient infrastructure, which is unacceptably detrimental to human well-being,

■ an economy that has ceased to be able to support the population's expectations for either "wealth creation" or "quality of life", and

■ a social environment that has become dysfunctional, resulting in increased stress and fear of crime, alienation, high crime rates, and subsequent outward migration.

(Ekins & Cooper, 1993)

Those that can afford it 'vote with their feet' and move out to the 'safer' and 'quieter' suburbs resulting in increased demand for transport, greater urban sprawl and less efficient suburban building forms (this is explored further in Chapter 4).

LAs are gradually bringing LA21 policies into action, so increasingly development has to take place in the context of, respond to and/or work within the sustainability agenda of the local area. In addition the UK Government has produced a discussion document as part of its strategy for more Sustainable Construction (DETR 1998). Although this acknowledges the 'wider agenda' it concentrates on the relative inefficiency of construction, particularly in terms of resource use, compared with other industrial sectors.

3.8 RESOURCE REDUCTION TARGETS FOR MORE SUSTAINABLE BUILDING DEVELOPMENT

The development, use and maintenance of buildings result in the consumption of 6 tonnes of building material/person/year in the UK (BRE 1996) and this is estimated to be 10 tonnes in the USA. Such figures bring home the implications of life to each citizen of the modern industrialised community as well as illustrating the very large quantity of resources that are consumed by construction and related industries. If the ethical dimension of sustainable development are accepted, i.e., in terms of equity, then material and component selection has to consider overall resource use, the limits to the stock(s) of raw materials and the access to the resources by future generations on the grounds of intergenerational equity. Consideration of international equity also means ensuring that less developed nations have the capacity to develop a similar standard of affluence as that enjoyed in the developed world through equality of access to the worlds' resources. If these objectives are to be met then a sustainable global society must seek very high resource efficiency gains. Such gains will be necessary to ensure that, as less developed nations industrialise their economies and increase their resource consumption, further irreversible damage to environmental systems is not caused and planet Earth can be maintained for future generations. Predictions of the resource efficiency gains that need to be achieved for society as a whole vary from Factor 4 (75% reductions) to Factor 20 (90% reductions). Such targets may seem "off the planet" from a current design and construction perspective but nevertheless many projects have shown some elements, such as energy or water efficiency gains, where factor 4 savings have been achieved by the application of currently available technologies (Weizsäker et al 1997). The Dutch government has instigated a research programme to explore how their construction industry can achieve Factor 20 by 2050. The recent BRE Environment Office is a good example of what can be achieved through careful and considered design. It achieves 30% energy reductions and has a 90% recycled content (BRE 2000-3). There are many other examples that could be cited, which collectively illustrate that if all were to be applied together, a Factor 4 gain in efficiency for the construction and use of new buildings is a feasible goal. Nevertheless this begs the question of how similar savings can be achieved in the use and maintenance of the vast majority of the existing building stock.

Targets such as these raise the question of the need to build and the location. Designers and their clients should consider seriously whether there is a real need for a new building. The main volume or mass of material lies in the structure and fabric so avoiding this resource expenditure through the reuse or conversion of an existing building is an important consideration at the feasibility stage. This is one way of extending the service life of the building and its components. Clearly longer life will mean less resource requirements, and this is an important factor in material and component selection in both refurbishment and new build design (see Chapter 4) and is a factor that is considered in the data sheets in Chapter 6. If a need for a new building can be demonstrated then resource reduction targets mentioned above (i.e. 75% reduction overall) are an essential requirement. There are a number of social and economic barriers to achieving such savings. Key problems include:

- The lack of demand by users for more resource efficient ways of living and working;

- Unequal or unfair distribution of benefits e.g., cost in use savings are made by users/occupiers and are not seen by developers/investors, who may be expected to fund additional capital or design costs;

- Additional design time required for refurbishment schemes and to locate and procure recycled materials and components;

- Perceptions that recycled materials are less durable or of inferior performance. This may be a real problem in that it is often difficult to ensure that recycled material meets British or International Standard requirements and may mean that specification should be examined to check whether high quality performance is required in all circumstances.

3.9 TOWARDS SUSTAINABLE BUILDING DESIGN

How then can the multiple objectives of increasing resource and energy efficiency, reducing negative economic and social impacts, and improving the local and global environments be achieved? Table 3.3 identifies a range of actions that should be taken by building designers and specifiers.

Figure 3.3 **Actions in pursuit of sustainable building**

- Review the need to build and if this is the case explore whether refurbishment of an existing building is an option that meets the need at greater resource efficiency, before deciding on new build.

- Audit and justify all the resource inputs to the proposed building and its development process as a whole including land, energy and water management strategies. Key to this is to select building materials and components of low energy and resource intensity

- Reduce dependency on central infrastructure, including water, energy, solid waste and human effluent.

■ Minimise waste and pollution outputs to air, land and water, over the buildings life cycle, including the manufacture, construction and demolition stages. (For the reasons identified, at present it is very difficult to undertake LCA in any systematic manner in the complete and entire meaning of the words, but we should set this as a longer term objective. Analysis of aspects such as energy use over the life cycle is just about feasible at the current level of understanding and at the current stage of development of tools for analysis.)

■ Consider life cycle costing and identify whole building and sub-component lifetimes at the design stage, with appropriate strategies to facilitate construction, refurbishment, dismantling, re-use, recycling and disposal. (i.e. design for disassembly, reuse and recycling.)

■ Replace flora and habitat for fauna destroyed or displaced by development (also consider issues such as deliberate reintroduction of locally relevant species, planting drought resistance species in water stressed areas, etc.).

■ Provide for local waste management, local storage and recycling together with provision for local human waste-water and effluent treatment.

■ Consider access to all facilities by all sectors of the community.

■ Consider building and landscape relationships that improve the sense of security and help reduce crime.

■ Ensure provision for local community participation in decision making in terms of how the proposed building will impact on the human development aspects of local Agenda 21. E.g. through citizens advisory committees, action groups etc.. This includes understanding the effects upon local employment, use of local materials, products, components and technologies, fair trade and fair employment practices, access to mobility and to local resources.

A number of these issues are relevant to the performance of the whole building and are returned to in chapter 4. Although it is clear that construction, use and maintenance of buildings and the associated consumption of building material resources is a vital aspect of future moves to make communities more sustainable, broader sustainable development of the wider community is beyond the scope of this book. So it is important to realise that this objective cannot be addressed purely through the design of individual buildings and material selection. Although good practice in these areas are an important contributory factor, many of the broader sustainable development issues are controlled by decisions taken at the planning, urban design and economic development level, with social and political consequences outside the direct control of the building designer and specifier. Such issues include land use efficiency, the location of the building(s) and availability of transport systems, the influence of the development on the economic sustainability of neighbouring areas, etc. Nevertheless these dimensions are very important to the developer and should be kept in mind at the inception and feasibility stages of projects.

Taken together all these points illustrate very clearly that the built environment and how we use and adapt it over the first half of the 21st Century will be crucial to the search for more sustainable development. All construction professionals need to become much more aware of this wider agenda.

Samantha Nicholson is an environmental consultant at Groundwork Trust in Manchester.

3.10 REFERENCES

Baldwin, R., Yates, A., Howard, N and Rao, S. (1998) BREEAM 98 for Offices: an Environmental Assessment Method for Office Buildings. Construction Research Communications Ltd., Watford, UK.

BEES (Building for Environmental and Economic Sustainability) (1998) Technical Manual and User Guide, National Inst. of Standards and Technology, 6144 Gaithersberg, MD, USA.

BEQUEST (1999) Information Papers: http://www.surveying.salford.ac.uk/bqextra

Building Research Establishment (1996) Buildings and Sustainable Development, Information Sheet A1, BRE, Garston, Watford, UK.

Building Research Establishment (2000 – 1) Environmental Profiles of Construction Materials and Components: http://collaborate.bre.co.uk/envprofiles

Building Research Establishment (2000 –2) ENVEST: http://www.bre.co.uk/envest

Building Research Establishment (2000–3) BREEAM: http://products.bre.co.uk/breeam/breeam4.html

Cole, R.J. and Larson, N., (1997) Green Building Challenge Assessment Manual, Vols 1 & 2, CANMET, Natural Resources Canada, Ottowa, Canada

Cooper, I., 1997, Environmental assessment methods for use at the building and city scales: constructing bridges or identifying common ground? in Brandon, P., Lombardi, P and Bentivegna, V. (eds.), Evaluation of the built environment for sustainability, conference proceedings, Florence, Italy, 13-15 September, 1995, E&FN Spon, London, Pgs 1-5.

Curwell, S., Yates, A., Howard, N., Bordass, B., and Doggart, J., (1999) The Green Building Challenge in the UK, Building Research and Information, Vol. 27, No's 4/5.

Department of the Environment (1994) Sustainable Development: the UK strategy, DOE, London.

Department of the Environment, 1996, Indicators of Sustainable Development for the UK, HMSO, London.

Department of the Environment, Transport and the Regions (1998) Sustainable Development Opportunities for Change – Sustainable Construction, DETR, London.

Department of the Environment, Transport and the Regions (2000) http://www.environment.detr.gov.uk./ga/change.htm

Ekins, P. & Cooper, I., (1993) Cities and sustainability, a joint research agenda for the Economic and Social Research Council and Science and Engineering Research Council, SERC, Swindon, England.

Friends of the Earth Scotland, (1995), draft, Strategy for a Sustainable Scotland, FoE Scotland, Edinburgh.

Forest Stewardship Council (2000) http://www.fsc-uk.demon.co.uk

International Tropical Timber Organisation (2000) Technical Report: Review of Progress toward the Year 2000 Objective. http://www.itto.or.jp

Mitchell, G., May, A., and McDonald, A. (1995) PICABUE: a methodological framework for the development of indicators of sustainable development, International Journal of Sustainable Development and World Ecology, Vol. 2, pp. 104-123.

Official Journal of the European Union (1998) No. 286

Palmer, J., Cooper, I., & van der Vorst, R. (1997) Mapping out fuzzy buzzwords – who sits where on sustainability and sustainable development, Sustainable Development, vol. 5, Issue 2, Pgs 87-93.

Rotheroe, C et al (1997) "Do the indicators of sustainable development produced by the UK Government and indicators developed within various local Agenda 21 initiatives have common characteristics from which core indicators can be developed?" Proceedings of the 1997 International Sustainable Development Research Conference, Manchester, pp 238-245, ERP Environment, UK.

United Nations Food and Agriculture Organisation (1997) State of the World's Forests - Executive Summary. http://www.fao.org/forestry

Stratospheric Ozone Research Group (1996) Report – Executive Summary, Department of the Environment, Transport and the Regions. London

United Nations Association (1995) Towards local sustainability: a review of current activity on Local Agenda 21 in the UK, UNA-UK, Sustainable Development Unit, and the Community Development Foundation, London, England.

United Nations Conference on Environment and Development (1992) The Earth Summit '92, Regency Press, London.

United Nations Environment Programme (1987) Montreal Protocol on Substances that Deplete the Ozone Layer, Final Act.

United States Environmental Protection Agency (1988) Environmental News Press Release, September 26.

Vale, B. and Vale, R. (1991) Green Architecture: Design for a Sustainable Future. Thames and Hudson Ltd., London.

Von Weiszaker, E., Lovins, A.B. and Lovins, L.H. (1997) Factor Four: Doubling Wealth, Halving Resource Use. Earthscan Publications Ltd., London.

World Commission on Environment and Development (1987) Our Common Future. Oxford University Press: Oxford and New York

CHAPTER FOUR

Whole Building Performance Issues – Design, Construction, Use and Disassembly

S.R.CURWELL, BSc MSc ARIBA

4.1 INTRODUCTION

Selection of 'healthy', 'green' or environmentally friendly building materials form one important aspect of a 'green' building. This chapter explores some of the broader green or environmentally friendly building performance issues. Material and component selection, using health and environmental criteria, cannot be addressed purely at the detailed design stage or as a simple detailed design issue because decisions on the form of the building can be a major determining factor. For example, the choice between a pitched or flat roof form of construction. The options for the structural materials and connections, the waterproofing system (slates, tiles, troughed metal sheeting or bitumen membrane) and possibly the type of insulation will all be 'predetermined' to some extent by a decision to use one roof form rather than the other.
In many cases this decision will be intuitive – emerging from a subconscious analysis of the interplay of function, form, location and aesthetic considerations. Alternatively the form and material choice may be a constraint imposed externally, e.g., by planning requirements.
Therefore the chapter supplies a range of good practice advice for the design, construction and maintenance of houses. However in a reasonable space it is not feasible to address all possible design and performance issues. Nevertheless the chapter attempts to relate the environmental and health performance of material and components to that of the whole building.
The importance of overall resource conservation and the fact that many technologies that enable radical reductions of 50% or more in energy and water consumption or material content of new construction was highlighted in chapter 3. The aim is to introduce designers and specifiers to some of the main options and to provide a qualitative understanding of the relative impacts of a range of technological solutions.

4.2 ADAPTABILITY AND THE LIFE EXPECTANCY OF BUILDINGS

Clearly one way to significantly reduce environmental impact and contribute to more sustainable building is to reduce resource consumption by ensuring that buildings have a long life as well as through effective dismantling in order to reuse and recycle the construction materials at the end of a building's life. The quality of design and construction will have considerable influence over the life expectancy of a building. Building obsolescence is a complex subject involving the interaction of a wide range of functional, technical, social, cultural and economic factors. Clearly if a building is of high quality and functions well in all respects and is adaptable to the changing lifestyles of users and of the community in which it is located, then there will be much less risk that its own life will be short. From an environmental perspective the most undesirable outcome is a short life for a building made using large quantities of energy intensive non-renewable materials, coupled with most of the subsequent demolition waste going to landfill. In this context the term 'waste' is a misnomer.

The UK housing stock as a whole is very long lived, the current rate of replacement is in the order of 100 years, however this conceals the fact that certain dwelling forms, such as some of the system built high rise, have been replaced in much shorter time-frames. Most dwellings are designed with a life expectancy of sixty years, but this is a vague target and really refers to the structural system and the main elements of the fabric. Occupier's living patterns and expectations will change considerably in the life time of a typical dwelling, so that it is likely that changes will be sought to the interior at least, even if they may not be required from a technical perspective. The heating system will require major overhaul at least once, probably twice, similarly the bathroom and kitchen equipment two or three times in 60 years. Changes in fashion may result in these latter items being changed more frequently. Thus parts of a building such as the heating system, interior fittings and finishes will be refurbished and/or replaced much more frequently, in a cycle that can be as short as 5 years. It is important to appreciate that a decision to replace may be influenced as much by aesthetic as by functional or technical considerations. Design for adaptability is a complex topic worthy of further detailed research and development in the context of extending the life of whole buildings and sub-components. Design for ease of maintenance, including dismantling and replacement of parts is also an important factor, which to date has been given relatively little attention in the construction industry as a whole. A small proportion of buildings are demolished prematurely, at least in terms of the normal life expectancy of the majority of the structural components, precisely because of the expense of maintenance to some ill-designed, ill-considered, poorly built or in-accessible sub-component part.

In terms of whole buildings, in principle, two main options can be identified. One is the long life building constructed from durable and heavy materials in a form that permits maximum adaptability over time. A good example is the various forms of the brick built terrace (Georgian, Victorian and Edwardian). Although it is accepted that the original developers of such buildings did not necessarily see them as long life buildings, the very long life of the structure of some of these buildings is related to a combination of factors. Clearly the fact that many contribute to the cultural heritage of the towns and cities which they form a part is a very important factor in their survival. Nevertheless their adaptability for dwellings of various sizes including sub-division into flats and for a range of small office and commercial uses, including incorporation of building services and communication technology as it evolved has been most important. The long life of some of these structures justifies the initial investment of the human, material and energy resources that went into their manufacture.

Alternatively, at the opposite end of the spectrum, one can envisage a short life building designed to be a close fit to the functional requirements, made from prefabricated and dismantle-able lightweight materials so that the parts can be readily and easily reconfigured for a new use, perhaps many times. The British Pavilion for the 1993 Saville Expo is an example of this approach. Some of the steel 'Case-Study' houses built in California in the 1950's and '60's by Charles Eames and other designers also had this potential. The Expo pavilion was subsequently dismantled but still awaits a new use and reconstruction. The investment in materials and energy will be returned in the longer term, provided that the parts see a long service life even though the original building use application has not. This short-life, tight-fit, easily reconfigured/relocated option is further into the future requiring more research and development, particularly in terms of dismantle-able foundations and the provision and relationship with utility services, although the technology for flexible reconnection to services (water, drainage, etc.) has long existed in the holiday park industry. Such concepts offer the possibility of 'minimum environmental disturbance' returning a site to its original use after the building is removed. Current housing concepts in the UK and elsewhere lean much more towards the longer life option, despite timber and light steel-framed structural forms of volumetric prefabrication being a reality. Steel systems are extensively utilised in the offshore and temporary 'cabins' industrial sectors and for holiday and mobile homes. The aesthetic images created by the current application of such technologies may seem entirely inappropriate to the traditional concept of 'home' and thus to many current householders. However if some of the predictions of societal change in the new age of information and communication technologies become a reality, such as the more transient nature of work and leisure, then it will only require the application of creative re-design and development to make new forms of super-adaptable dwelling. Because of the super-adaptability provided by the reuse/reconfiguration design possibilities of these technologies such forms appear to offer the potential of lower environmental impact over time than that likely to result from the need to make very frequent and radical changes to current long life housing forms.

Obviously other housing forms may be developed occupying intermediate positions between these theoretical extremes, as citizens are likely to find compromises between the radical change implied by very transient use patterns and the need for cultural heritage and continuity. Nevertheless it is now very clear that the current assumptions made by client, designer and component manufacturer regarding whole house and sub-component life expectancy must be questioned in the context of the emerging concepts of environmental responsibility and sustainable development.

4.3 LIFE-STYLE AND EXPECTATIONS OF BUILDING USERS

Lifestyle choices and affluence have a considerable impact on dwelling forms. In most countries higher affluence leads to greater expectations for space and facilities e.g. en-suite bathrooms, home office, workshops, swimming pool, etc. In the USA and Australia and to some degree in the UK larger dwellings usually occupy much larger plots. Housing is supplied in single family units of single and two-storey form, gardens are larger, all located in suburban areas. In all countries affluence also leads to ownership of more property such as a main dwelling and a holiday home. This all adds to the increasing problems of ubanisation. Containing urban sprawl is an important issue for the future development of towns and cities. In addition to the loss of land and green space urban sprawl results in greater demand for transport services and the higher resources to create the associated utility infrastructure. Therefore increasing space and land use efficiency whist still meeting users needs is a design priority (DETR 1999). Also smaller houses use less construction materials and energy.

In certain European countries there is a long history of and a strong collectivist tradition in apartment living. There are a number of interacting socio-economic and cultural reasons for this but it is widely recognised that the medium rise apartment block offers greater urban density and overall efficiency particularly in terms of transport. Many also consider that the overall urban design configurations using this form also offer improved quality of life. (quote RR – Intro to Urban task force report). The use of apartments and maisonettes in larger buildings offer many economies of scale over single family houses in terms of lower overall surface area, lower heat loss and heating requirements and more efficient provision of utility services. The greater community spirit and sense of collectivism engendered in this form of housing provision can lead to other environmental and resource reduction benefits resulting from sharing of facilities and services, e.g. laundry, recycling, composting, gardening for food production and car-pooling and sharing. Greater consumer environmental awareness is leading to additional need and expectation for storage space for waste, recycling and composting facilities in all housing forms.

Space for a garden is considered important by a majority, and to grow food by a sizeable minority, of UK citizens. Both can contribute to a healthier life style and as modern agriculture and food production is energy and chemically intensive can reduce environmental stress. Space to grow one's own food is advantageous provided it is not at the cost of a higher land take, which can be ameliorated through the provision of roof gardens and/or allotments.

Public perceptions of the quality of life in the city and countryside influences the place selected to be 'home' and 'work' and form key decisions for each citizen, having considerable effects on transport requirements. Social problems such as the perception of high crime rate as well as pollution problems in inner cities can lead to outward migration by all those who can afford it and contribute to upward spiralling trends in car ownership and use especially where employment and dwelling remain zoned in different sectors of the city. This is despite the nature and quality of the public transport provision, as there is an increasing tendency for the affluent to see local public transport only for the use of the 'poor', the young and the aged.

User expectations of the performance of buildings in terms of environmental control and comfort are becoming more demanding. From a design perspective there are some conflicting expectations which present difficult challenges. Smoking is a key example. In public areas of buildings the best method is control at source, i.e., to restrict smoking to designated areas with separate ventilation systems. Within individual dwellings such an approach challenges freedom of choice, but smokers should be aware of the risks of passive smoking to other members of their household and ensure high rates of ventilation (10 ach.) to disperse the pollutants. Most users health concerns identify the requirement for 'fresh air' which is potentially in conflict with the need to restrict ventilation rates in winter conditions in order to provide thermal comfort without excessive energy consumption. Allergy sufferers emphasise the need for a completely 'chemically' free environment, which is difficult to supply with the current range of building materials, although progress in terms of reducing off-gassing from some products has been made in recent years.

Finally many users have come to expect responsive control, e.g., for a cold room to be brought up to temperature quickly and vice-versa almost at the 'click of a switch'. Passive solar and ventilation strategies when coupled to a very well insulated envelope have the potential to reduce energy requirements to a very low level when functioning properly. In fact the zero space-heating house is a technical reality. Sophisticated weather compensated control is now available to the domestic consumer. So increasingly more responsive 'fingertip' control is becoming available at higher capital cost and possibly compromising overall energy efficiency if users are unable to 'operate' the house systems in an intelligent manner over time. Thus users may have to adjust their life-style slightly and/or understand how the building can be operated to best advantage. Some users may not be prepared, or may be unable to do this. Therefore understanding potential user's perceptions of these matters, and other issues, such as the perception of risk and hazard is important at the briefing stage in order to ensure compatibility with user expectations of building performance.

4.4 SITE AND LOCATION ISSUES

Location can have considerable influence on a number of health and environmental issues. Orientation, topography, the nature of the local vegetation and form of buildings are core design criteria in terms of the provision of high quality dwellings that have a beneficial effect upon the well being of users through access to sunlight, air, good aspect and prospect, etc. They also determine the solar energy available, the local wind speed and air quality. A north-facing slope may offer considerably less solar gain than that of a south-facing slope. The bottom of deep valley may be in shadow for a considerable period of the day, as will the deep street canyon of the modern high-rise city. Air pollution may also be concentrated in these places. It is important to realise that addressing these issues early in the design process is essential in order to maximise the advantages and minimise the disadvantages of any particular site location. For most places optimisation of these factors will be more important to the provision of a 'healthy' building than the selection of building materials. Also it will usually be equally, or more, influential in reducing the environmental impact over the overall life of the building, e.g. by reducing energy use though passive solar heating when compared with the selection of low embodied energy materials. Ignorance of these factors in design can also result in buildings of poor overall energy efficiency due to the their inability to make effective use of the 'free' solar gain or through the need to provide mechanical cooling due to excessive overheating in summer.

Pollution and noise emanating from local industry or from road and air transport is best prevented or ameliorated at source. This is essential if the type of mixed development considered important to the transition to more sustainable communities is to be a possibility. Spatial separation from the source is an alternative but this implies either zoning of industrial usage away from housing and/or areas of land that cannot be brought into productive building use, although they are useful as 'green' space. Steps to handle noise and pollution by the design and construction of the building will cause excessive costs, e.g., by sealed windows and mechanical ventilation, and may not completely cure the problem. Electo-magnetic fields from overhead power lines and electrical sub stations have been implicated in some forms of cancer such as leukaemia in children but at present no clear explanation for this has been identified.

Inner city 'brown-land' sites have the benefit of recycling unused and/or under utilised space and of minimising the requirement for new utility services and transport infrastructure. However contamination of brown-land sites due to previous industrial use is a major environmental and health problem in many of the post-industrial areas of UK cities, with serious economic consequences that undermine and negate efforts to bring this land back into 'productive' use. Grants are available, e.g. derelict land grant, in order to facilitate clean up but this may not address all of the direct and indirect costs. A key indirect cost is the reduction in value of the property due to potential purchaser's negative perception of the risks. This is real as the risk to health can far exceed that from other sources so it is important that expert advice is sought on contaminated sites in order to effectively rehabilitate the land. Unethical and unscrupulous housing development on inadequately treated contaminated sites has damaged consumer confidence and this now forms an additional barrier to redevelopment of parts of the inner city.

In larger housing schemes the provision of public transport to the finished scheme should be a priority and the effect of the increased car use caused by the proposals addressed both in terms of local pollution and congestion, but also with the local community. The tendency to design current housing layouts around road/car use should be challenged and walking/cycling/public transport options used as the primary generator for planning.

Transport of the materials and components to the site during the construction phase is another consideration. In the majority of locations local materials and components are to be preferred both in terms of reducing pollution from transport as well as for harmonisation with the built forms and materials of the area. Transport pollution is a more important consideration for the heavy bulk materials such as concrete, aggregates, bricks, tiles etc. If materials need to be brought from further afield then lighter materials should be considered to reduce the weight to be transported. In principle with very inaccessible locations very lightweight materials such as light steel or wood frame should be considered, but often in the UK such locations are precisely the areas of outstanding natural beauty where (heavyweight?) vernacular solutions are required or are most appropriate to harmonise with the locale. In all cases suppliers should be encouraged to co-operate to reduce part load inefficiencies and site personnel encouraged to car pool. Dust from vehicles accessing sites needs to be controlled in order to reduce nuisance.

4.5 FUNCTION, FORM AND LAYOUT OF DWELLINGS

In theory the cube provides the maximum volume to surface area ratio (not considering curved forms) which in principle will reduce material use in, and energy loss through, the fabric. On the other hand day-lighting and natural ventilation, which also have the potential to reduce energy consumption and are preferred by most users, are normally facilitated by narrower plan forms such as the terrace. One of the reasons for the popularity of the single-family house and garden is the fact that the building design is non-critical. It is relatively easy to manipulate the design of individual houses or small terraces in order to create an effective compromise between efficient building form and the requirements of daylight, sunlight and natural ventilation to all rooms. It is also easy to extend into the garden to provide more space as and when required. However, as mentioned elsewhere, collectively as an urban form at the low densities normally achieved, single family housing provides much lower land use and lower overall resource efficiency than other more compact forms of housing provision.

Earth sheltering and or underground construction both provide ways of increasing the thermal efficiency of buildings and as a means of noise suppression. Earth sheltering is normally provided on the north side often using the beneficial inclination of a south-facing slope with the main aspect of the building facing south. Construction costs may rise due to the additional damp-proofing provision, but can be offset by the energy savings in use. This form may constrain extension and be more difficult to adapt to other uses.

In larger apartment buildings, there is a creative tension in design between the provision of an efficient building form and the provision of daylight, sunlight and natural ventilation to all rooms in all of the units. In deeper plan buildings a higher ceiling height, a minimum of 3.0m, is necessary if daylight and natural ventilation are to effectively penetrate the interior. This is higher than is strictly necessary for normal use, however the additional height provides greater adaptability in terms of the provision of new or revised building services in the future, as well as greater possibilities of change to other non-domestic uses.

4.6 STRUCTURAL ALTERNATIVES

Small low-rise single family houses allow the application of relatively simple building technologies. It is possible to assemble 'by hand' without the need for sophisticated equipment and cranes. With timber technology assembly is very easy with semi-skilled labour (e.g. self-builders). Larger apartment buildings require more sophisticated techniques such as reinforced concrete, concrete stairs, lifts, etc., which require cranes and skilled labour to assemble. Most of the discussion below concentrates on low-rise single family housing.

Small buildings use simple spread foundations. This requires efficient design to reduce concrete volume and waste. Timber piles are used in USA and could be considered in the UK. Durable timber must be used otherwise chemical preservatives are required to prevent decay. With larger buildings concrete piles may offer the possibility of reuse for a new building on the old piles at end of life of the original building founded upon them. (eg new office building in Manhattan - Kaplan 1997)

Three common alternative structural forms are used for single family houses in the UK: load-bearing, timber and steel framed. Timber and steel framed systems are usually employed with brick veneer cladding, so the comments below are based on the assumption of brick veneer. Bricks for the outer leaf/cladding are preferred by many householders and offer very good durability and performance in UK conditions, usually justifying the initial energy and resource investment in their manufacture. Some manufacturers are making progress in reducing the environmental impacts of brick making such as using methane harvested from domestic waste to fire brick kilns. Clearly this is advantageous over other fossil fuel sources. Framed systems can be clad in timber and other sheet materials such as fibre-cement sheet in which case these options have the advantage of lower initial embodied energy over normal North Sea gas fired bricks. Given the common consumer preference for brick, then, in terms of the common technologies currently applied in the U.K., the real comparison is between concrete blocks, steel and timber as the main load-bearing element within the brick outer cladding. In this context timber and steel have to be considered with the enclosing and lining elements, normally plasterboard internally and plywood externally, which means that the decision on the lowest environmental impact is not clear-cut.

Timber is a renewable material of low embodied energy. The vast majority is imported into UK therefore transport energy becomes critical in terms of embodied energy comparisons with alternatives. It is important to ensure that timber comes from sustainably managed sources – see data sheet. It is easy to install high levels of thermal insulation in framed systems and there are many advantages in terms of buildability using the common platform frame construction method, when compared to load-bearing solutions. There is still consumer resistance, such as concerns over fire, decay and condensation, but these are unfounded. A significant disadvantage is lack of mass, which increases the risk of summer overheating. Attention should be given to window design, size, orientation and to shading to minimise this problem. Timber is recyclable and possibly whole framed panels with careful design for disassembly.

Lightweight galvanised steel sections are easy to assemble with self-drilling, self-tapping screw technology in a similar platform frame configuration to wood frames, offering very similar buildability and insulation installation advantages over the load-bearing alternative. Steel suffers from similar lack of mass to wood frame but is easily recycled and/or reused. Steel has higher embodied energy, but as it is cost competitive with timber in UK market conditions, the greater reuse/recycling potential at end of life may justify higher initial energy investment over timber.

Load-bearing brick + block is seen as traditional solution although current cavity wall construction has little technology in common with older solid wall construction of earlier times. Lime mortars have the advantage of lower strength, which facilitates dismantling and reuse at end of life, although curing time at construction is extended. Weak cement/lime mortars are a good compromise to accelerate curing. From the 1970s very lightweight aerated concrete blocks became a standard solution as part of the thermal insulation system. These blocks manufactured from power station waste (PFA – see application sheet 7.3) offer little thermal mass and have relatively high embodied energy. Understanding the role of the building mass in passive control of summer overheating should increase use of dense blocks with partial or fully filled cavity insulation. Very wide cavity construction using dense blocks and full-fill insulation in cavities offers super-insulation at the cost wider walls (250mm+), but with excellent overall thermal control and performance.

Where short life is anticipated it is clearly better to avoid high embodied energy materials that are difficult to dismantle and reuse, such as bricks and blocks using strong cement mortars.

4.7 DETAILED DESIGN

Life expectancy of sub components must be considered at the design stage so that adequate access can be planned at the outset for repair and replacement. For example a balcony can have multiple functions, as well as giving access to outside windows it can also function as a shading device for a window below and to provide access for periodic painting of timber cladding above. Design of access to services in floors, walls and ceiling voids is very important both in terms of replacing sub components and also for ease of maintenance. Surface mounting of pipe-work and electrical wiring can add flexibility and has the added advantage in reducing access to concealed spaces for a number of pests of buildings, insects and vermin. Consideration of cleaning requirements is also important in the latter context. House mites have found a comfortable home in the soft furnishings and temperature and humidity levels of our houses and have been implicated in the epidemic of asthma in children in the UK. Using hard finishes is one way to reduce their 'habitat', but present acoustic problems.

In terms of provision for dismantling and disassembly the assumption that prefabricated 'dry' components and assembly are better needs analysis. Simple issues may be more important, such as using weak mortar compositions for 'wet' assembly. In principle mechanical fixing (screws and bolts) are likely to be better than adhesives from the standpoint of disassembly, particularly in terms of disassembling larger components. An effective market for salvaged reusable building materials is gradually becoming established in the UK. Sanitary ware and architectural features have been sought and salvaged for many years. Slates, stone, older hand made bricks and timber are now routinely recovered. The DETR's materials exchange may assist in the market making in re-useable and recyclable materials. Recent developments such as the land fill levy are likely to provide a greater incentive to recover virtually all the main structural material even if it is simply crushed for hardcore. At the very least this reduces demand for crushed rock or natural gravel from other sources. The recent Environment Office constructed at BRE achieved 90% recycled content and although this is an unrealistic routine target for every project at present it indicates what can be achieved. The organisational and cultural barriers to increased recycling need to be overcome. Designers should seek to maximise the recycled content. Commitment to this principle requires greater lead time and planning as well as partnerships with demolition contractors in order to strengthen the market in recovered building materials. Demolition contractors need to establish effective inventory information for use by design teams.

For both new and recycled material and component selection the use of local sources in terms of ameliorating transport impacts has been emphasised previously. The data sheets in Chapter 6 provide considerable guidance for selection on technical, health and environmental grounds and so this is not explored further here. In principle if a 'safer' or 'greener' material is available at an acceptable or similar cost, that satisfies technical requirements, then use it! Organic materials, such as timber, have the advantage that they are renewable – in principle we can grow more. However organic materials are susceptible to decay. In the UK the main agent in this is dampness, and so in order to achieve the best performance from timber and other organic materials it is essential to adopt good building construction and maintenance practice to maintain dry conditions in the interior. The use of timber preservatives should be seen as a last resort in areas where there is known wood boring insect infestation or this can be anticipated (e.g. termites in hot climatic zones), or for those components in very exposed locations, such as softwood windows and cladding. Dry conditions are also important in reducing the risk of damage through frost action in inorganic materials. The effect of climate change on building materials will also become an issue in the future in terms of the risk of greater rainfall, temperature and wind speed fluctuations.

4.8 HEATING, VENTILATION AND BUILDING SERVICES

The overall energy efficiency and level of comfort achieved in dwellings are determined by the interaction of a number of factors:

1. The thermal performance of the fabric as a whole, i.e. including the walls, roof, floors, windows and doors,

2. The ventilation system (passive systems, i.e. by opening windows, or active systems using mechanical ventilation),

3. The airtightness of the fabric, including the windows and doors as well as the main constituents of the walls and roof,

4. The efficiency of the heating system, i.e. in terms of the boiler efficiency, efficiency of pumps, etc.,

5. The effectiveness of control systems and users in tandem in optimising the overall performance to the seasonal and weather variations,

6. The operating costs – fuel prices and maintenance costs.

Building Fabric

Experiments in the early 1980s and subsequent experience in a number of buildings has indicated that it is possible to super-insulate the fabric and to virtually eliminate the need for space heating in the UK's temperate climatic conditions. Reductions of 70%+ in space heating below that of buildings constructed to recent standards are quite feasible and should be actively considered as a minimum target in all new construction (Webster 1987). In order to achieve this careful attention to orientation and windows is required in combination with insulation thicknesses of the order of 170-200mm in walls and roof and by increasing the thermal mass of the dwelling, e.g. by using dense concrete blocks for the inner leaf of cavity wall construction. The latter help to 'iron-out' diurnal and seasonal swings in outside temperatures. In effect super-insulation (a term used in this context to represent the sum of these measures) has the main advantage of reducing the heating season. That is the period when the users feel the need for heating from additional sources other than that available from solar energy and the incidental heat gains available within the dwelling (e.g. from cooking and from the sentient heat given off by the occupants). In fact it is possible to eliminate the need for central heating and its cost, by using the money on a higher standard of insulation! The energy required can be provided from a single source, such as a small gas fire in the living room, provided users accept that the responsive 'fingertip' control mentioned earlier is not available in other rooms. Currently most users do not consider this acceptable. If this were the case the additional cost of insulation could be seen as a simple trade off with the cost of central heating.

In order to maximise efficiency it is necessary to give careful consideration to orientation and the design of the windows and glazing in order to maximise the 'solar collector' effect. High specification double-glazing with U value of $1:0$ W/m^2/^{0}C provides a good cost–benefit factor in reducing heat loss in winter. Depending on site constraints the design of the house is usually best optimised with the main living rooms and bedrooms with larger windows orientated to the south or south-west and with the kitchen and bathroom with smaller windows optimised to provide adequate daylight to the north and east. It is not necessary to have 100% glazing to the south unless an attempt is being made to achieve energy autonomy (or to achieve the aesthetic effect of a 'green' building!). In fact the main danger with a super-insulated fabric and larger areas of glazing is the risk of serious overheating in summer. Solar control coatings applied to the glazing can reduce the risk, but probably the best means is through shading of the south facing windows. This can be achieved through horizontal shelves, balconies or roof overhangs or through the use of well-placed vines and shrubs or via various combinations of these. The main object is to exclude the high asimuth sun in summer whilst admitting the low asimuth sun in the early morning in the spring and the autumn and for the whole day in the winter. In this context deciduous vines that shed their leaves in winter are necessary. Using the combination of these features good optimisation of sun path and orientation for the various seasonal conditions, winter, summer and spring/autumn is relatively simple for single family housing due to the flexibility of layout on the plot, but is more difficult in larger apartment buildings in urban areas. However optimisation is possible with larger buildings using narrow plan 'slab' blocks and/or through the use of courtyards or atria, but this may be at the cost of larger surface area of external fabric.

Ventilation System

Ventilation is required to control odour and humidity, to supply adequate fresh air (oxygen) for the occupants, to create air movement required for optimal occupant thermal comfort and for effective combustion in heating appliances. The latter is a vitally important health consideration. There are still an unacceptable number of fatalities as a consequence of carbon monoxide poisoning through badly installed and maintained heating appliances and flues, so it is important to ensure that regulations for the supply of combustion air to heating appliances and proper maintenance schedules are complied with. Relatively, this problem forms a much greater risk to the health of occupants than that from 'hazardous' building materials. Gas appliances utilising through the wall 'balanced-flue' systems, which are sealed off from internal air, offer thermal efficiency advantages by reducing the excessive air change rates to the interior through normal 'open-flue' chimney systems. All flues need to be sited carefully to avoid cross contamination via opening windows.

Dampness and excess moisture continue to be a major cause of ill health in buildings. The main problems result from substandard dwellings, which are inevitably occupied by the poor and/or aged who can ill afford the cost of energy for effective heating causing additional problems due to condensation. Warm conditions with excess humidity (65%+) provides ideal habitat for many insect pests, e.g. house dust mites. As mentioned in the example in the Introduction (Chapter 1:5) dust mites have been implicated in the epidemic of asthma in the UK. Regulations for new buildings require removal of water vapour at source by passive stack or mechanical extract systems from kitchen, bath and shower rooms. Good design and construction practice to avoid dampness is necessary, as is effective maintenance to correct leaks from building services. In extreme cases of high humidity it may well be that de-humidifiers could be used to redress this problem.

One air change per hour is advised as the best compromise in terms of thermal efficiency, health and indoor air quality, **provided** that occupants do not smoke. Regulations require minimum ventilation openings (usually provided by slot ventilators in windows) in order to achieve this figure through passive air movement. Passive systems will result in some heat loss in UK winter conditions. Clearly higher rates, such as that resulting when a window is inadvertently left open in high wind conditions, will result in considerable additional energy consumption for space heating in the heating season. Mechanical ventilation can provide better control and energy efficiency. Small-scale heat recovery equipment is available and offers further energy reductions although cost-benefit factors mean long pay back periods.

For most occupants indoor air quality is important to comfort and wellbeing. The rate of ventilation is a key contributor to indoor air quality, but should not be seen as the primary means of control of indoor pollutants, which are better controlled 'at source'. Tobacco smoke is the most important pollutant of indoor air and presents greater risk to the health of occupants than a number of other potential sources. Concerns have been expressed over off-gassing and release of fibres from construction materials. Formaldehyde adhesives, solvents paints and wood preservatives as well as man made mineral fibres are routinely used in modern buildings. It is very important that loose fibrous materials are sealed within the construction – see detailed guidance on the data sheets for these materials. Progress in the development of adhesives in man made boards such as chipboard (which is a useful material utilising timber that would otherwise go to waste) and in reducing the solvent content of paints has reduced any hazard to occupants from off-gassing to a very low level. In new construction this can be further minimised by ensuring the building is well 'aired' before occupation. Nevertheless there are sensitive individuals where special precautions in material specification may be necessary in order to reduce the risk of an adverse reaction to the lowest level possible. There is greater risk during repair and maintenance when occupants should be decanted and high ventilation rates provided whilst work is in progress. Vigilance is still necessary in existing buildings as the risk from asbestos, old leaded paint and lead in water supply pipe-work continue to present greater risk to the health of the occupants than that posed by off-gassing from other building materials. Chapter 7 provides further guidance on dealing with existing buildings.

The air-tightness of the fabric

Excessive infiltration, i.e., unwanted ventilation through lack of air-tightness in the fabric needs to be minimised for two reasons. Firstly to avoid excessive heat loss with associated loss of energy efficiency and secondly because it can undermine the performance of both passive and active ventilation systems. In order to check for any serious leakage new buildings should be pressure-tested on completion. Concrete blocks are not airtight and so for load-bearing construction wet plastering and/or painting is important to the overall energy performance by ensuring air-tightness of interior block walls, although plaster impedes recycling of blocks at the end of the building's life. Modern sealant and gasket systems can ensure considerably improved air-tightness of doors and windows as well as sheet lining system and claddings. 'High performance' standard windows and doors should be specified and give good air sealing capability. Attention should be given to the life expectancy of sealants to ensure that the air-tightness performance is not impaired over time.

There may appear to be a potential conflict between the need for air-tightness as an energy efficiency measure and the need to ensure adequate ventilation for good health. This is not the case. The regulations and good practice guidelines identified above are intended to provide adequate ventilation for good health assuming the building is otherwise airtight. However care is necessary to ensure that the ventilation openings are not obstructed and to avoid defects in mechanical systems.

The improvements in air-tightness of buildings in recent years has resulted in a reduction in air change rates from 20+ach in rooms with open-flued chimneys to the 1ach expected of whole buildings of current construction. Although this means a beneficial improvement in energy efficiency it has exacerbated the risk from radon in a number of areas in the UK, including Cornwall, parts of Devon, Derbyshire and the Highlands of Scotland. Radon is a naturally occurring radio-active gas that enters the building from the ground below, which is known to cause increased risk of lung cancer. The risk to smokers is double that to non-smokers. Special precautions are necessary in these areas to seal the ground floor or to provide pressure relief below the ground floor. Householders can obtain a simple monitor from their local authority's public health office in order to check for excessive radon levels.

The efficiency of the heating system and other equipment

Specifiers should seek the lowest energy equipment, heaters, boilers, lamps and pumps. In well-insulated buildings the space-heating load is small and the main energy requirement is for domestic hot water. With the latter, instantaneous heating utilising a 'combined' boiler offers greater efficiency as it removes the need for, and heat loss from, domestic hot water storage, although the lower delivery flow of hot water in very cold weather is disliked by some users. (The heat from water storage is not totally lost as it passes into the interior and is often used to 'air' clothes). This system requires slightly less pipe-work. If central heating is to be installed in buildings with a small space heating requirement condensing boilers offer much greater part-load fuel efficiency (80%+) compared to conventional boilers (60%). These boilers cost 25-40% more but grants are available to many consumers (e.g. the elderly) to offset the capital cost disadvantage. Boilers over 10 years old should be replaced because progress in combustion and heating technology has led to much greater fuel efficiency in modern equipment.

Water collector solar panels can be effective in the UK with around 8-10 year pay back period for the commercially available systems using a pump, sensors and motorised valves to maximise energy yield. Simpler DIY 'gravity' circulation systems are less efficient but cost much less. Electric photo-voltaic systems are currently too expensive to justify return on investment for everyday use. Low energy florescent bulbs can provide savings of 75% over the much cheaper initial cost incandescent type.

Domestic equipment manufacturers have improved the energy efficiency of some of their refrigerators and washing machines by 25-30% over the last 5 years and so low energy equipment should be sought. Similarly washing machines are available using 50% less water than those of 5 years ago.

The effectiveness of control systems

Regardless of the thermal performance of the building fabric, control systems can make a very significant and effective contribution to reducing energy consumption. A building management system controlling features such as optimum start up, zoning and motorised valves together with individual thermostatic valves on radiators as well as control over artificial lighting can provide a very a close fit to use and occupation patterns thus avoiding waste. Further efficiency could be achieved utilising optimum start/weather compensation, which has only recently been developed for the domestic market, but the payback period is likely to be extended. As yet such a complete package of sophisticated control(s) is rarely fitted in houses. In addition to cost constraints, this may be related to demand for or the ability of certain householders to use such systems.

4.9 USE, OPERATION & ROUTINE MAINTENANCE

From a health perspective physical safety is most important, as there is greater risk from accidents in the home than from other hazards. Regulations require proper design and construction of stairs and handrails and for glazed doors. Householders can also reduce risk by avoiding or securing loose mats and rugs. Smoke alarms are a very cost-effective way of reducing risk of loss of life and serious injury in house fires.

Effective preventative maintenance programmes form a key part of good operational planning. Most householders are aware of steps required to maintain exterior joinery and for heating appliances. The latter has already been emphasised, as regular maintenance of boilers, flues and fires is vitally important to protect health as well as to maintain fuel efficiency. Cleaning is also an important health consideration, but not just in the way most householders consider it, i.e., in terms of preventing 'germs' and reducing risk of disease, but also in terms of the risk from some of the cleaning solutions and from the pests of buildings. Some of the more aggressive cleaning solutions available are poisons and need to be stored securely, inaccessible to children. Some solvent cleaning solutions can cause allergic reactions. The need to reduce habitat for a number of the pests of buildings has been explained so designers can assist in the cleaning task by reducing or even better eliminating inaccessible service areas and zones, which provide living and communication spaces for many household pests. This is better than having to use insecticides or other poisons as a means of eradication when an infestation occurs. In terms of the house dust mite vacuuming carpets regularly with a special fine-filter cleaner capable of capturing the mites and their dust (actually the faeces of the mite) is an effective part of a control strategy. It is important to note that ordinary cleaners not equipped with fine filters may make matters worse for asthma sufferers by blowing the dust into suspension in the internal air.

Le-Corbusier famously described the house as a 'machine for living in', however users and designers alike have never really come to see the dwelling as a machine, e.g., in the same way as the motor car. Advanced sophisticated features and controls in cars seem to be desirable where they add to comfort or are at least accepted when they reduce pollution or save fuel (or allow the use of a larger vehicle at modest fuel consumption!). Clear operation and maintenance instructions are expected of car manufacturers. This is not the case for houses. There are many historic, cultural and economic reasons for this. Current inadequate approaches to, and understanding of, the operation of dwellings needs to be thoroughly addressed, from the education of schoolchildren through to that of construction professionals, in order to improve energy efficiency, reduce waste and risk to health. As already described for the most efficient operation the majority of low energy, low environmental impact houses are likely to require optimisation of passive ventilation, with effective utilisation of the solar energy available and with the active heating system over each seasonal cycle. Therefore good operational manuals for dwellings must link the operation of passive features of the fabric, such as opening of windows to the control of the active building services and to the heating or cooling season. For example the importance of operation of ventilators and opening windows needs to be explained in terms of health, humidity control and for summer cooling and the need for the proper maintenance of windows that ensures their proper function, e.g., not becoming impaired by jamming by paint.

Ideally the systems should be simple and easy to understand and operate. This does not mean that complex servicing systems should not be used, it means that the control and operational systems have to be made easily understandable, should not have counter intuitive features and be useable by the majority of the population, including the elderly and infirm.

■   The author acknowledges the contribution to this chapter of David Dowdle, School of Construction and Property Management, University of Salford.

4.10 REFERENCES

Department of Environment, Transport and the Regions (1999) *Towards an urban renaissance: Final report of the urban taskforce,* London, HMSO.

Kaplan, D. (1997) Manhattan's Green Giant, *Environmental Design and Construction,* Pg 22-28, September 1997.

Webster, P.J. (1987) *The Salford Low Energy House: A Demonstration at Strawberry Hill, Salford.* Salford University Report for BRECSU. ED 179/59.

CHAPTER FIVE

Use of the Application Sheets

5.1 INTRODUCTION

The focus of the guide is housing and is intended for use by all those involved in the design, construction, maintenance and alteration of these buildings. To derive the most benefit from the guidance contained in the sheets, it is considered important for readers to familiarise themselves with the contents of Chapters 2,3, and 4.

5.2 LAYOUT OF THE SHEETS

Each sheet is concerned with the various materials and components that could be considered for use in specific application on or within the building, and adjoining siteworks. There are further sheets of generic materials to support the applications sheets and to provide information on applications where only one material would normally be used, e.g. Concrete in foundations. The layout of the sheets is designed to permit a large quantity of information to be provided in a concise format. It is important to appreciate that the grades provided are relative between the various alternatives for any particular application. Therefore cross comparison of grades between different applications should not be made.

The main part of the sheet comprises a column listing the alternative materials likely to be considered followed by columns showing comments on the technical, health, environment and cost implications respectively. This forms the basis from which the guidance notes are formed.

5.2.1 Technical Requirements

These include, where appropriate, a diagram illustrating a typical application. The technical requirements provide a brief outline of the various considerations a designer may wish to bear in mind for this particular application and are based on current building practice in the U.K.. Comments are also included on the possible means of degradation of the material and potential routes of contamination of the occupants, which are provided to understand the service conditions of the material and possible means of degradation, decay or contact with the occupants. This is important; for example, contamination of foodstuffs or water supply involves greater potential hazards to health.

5.2.2 Technical Comment

Within the resources and scope of the study, detail testing and analysis of material performance was impossible so reliance had to be placed upon available technical data. The technical assessment is therefore a comparison of the materials against the 'average' conditions normally expected for each application. Thus the comments are intended to assist the designer/specifier and **do not** provide a definitive ranking for all circumstances. Individual designers will need to apply their own judgements in specific design situations.

A simple 1 to 10 scale is used, 1 being the most satisfactory through to 10, unsatisfactory. It is assumed that all materials comply with the relevant British Standards and Agreement Board certificates. The authors attempted to standardise the scale with the 0-3 used elsewhere, but concluded that the 1-10 scale proved a way of recognising the subtle differences in performance between some alternatives.

5.2.3 Health Comment

The comment derive from the general guidance given in Chapter 2 and provide detailed consideration of the health risks, if any, of each material in each application. The risk ranking (A/B) is set out below. A fuller description of this is provided in Chapter 1. It must be stressed that the ranking is a comparison of risk of each material in the specific application and is not necessarily a ranking of the risk of this material across all applications.

Figure 5.1

	Potential hazard when in position	Potential hazard when chance of being disturbed
	A	**B**
Non reasonably expected	0	0
Slight impact/not qualified by research	1	1
Moderate impact	2	2
Unacceptable	3	3

5.2.4 Environmental Comment

The comments derive from the general guidance given in Chapter 3 and provide detailed consideration of the environmental impact, if any, of each material in each application. The impact ranking is set out below. A fuller description is provided in Chapter 1. Similarly as in health comments above, the ranking in each category is based upon the relative impact of the materials in the specific application and should not be used as a comparison of the same material used in a different application.

Figure 5.2

	Impact upstream	Impact during construction	Impact during the life of building	Impact downstream
	A	**B**	**C**	**D**
Non reasonably expected	0	0	0	0
Slight impact/not qualified by research	1	1	1	1
Moderate impact	2	2	2	2
Unacceptable	3	3	3	3

5.2.5 Cost comment

The cost ranking figures provide a comparative guide for each separate application. The material having the lowest unit cost is allocated a base rank of 100 and all other material costs are compared with this. The ranking figures are only to be used as an indicator as it is not always possible to compare like with like. For example, some materials are produced in fixed modular sizes whereas others provide greater flexibility.

It will be noted that in some cases, the letters N/A have been set against individual materials. This indicates that either the material or the cost information relating to the material is not available or that we have been unable to obtain the cost information for the particular material.

5.2.6 Guidance notes

The gradings and guidance represent recommendations of the authors based upon a review of current literature and understanding encapsulated in the technical, health, environmental and cost comments. No original research or testing has been undertaken.

It should be stressed that these guidance notes are a suggestion only. Any individual making a design judgement must consider the problem as it occurs in that situation with their perception of risk and impact in conjunction with all other parameters effecting the decision.

The authors have spent considerable effort in reviewing information available and believe these guidelines to be valid. However, the authors would be grateful for any other comments and opinions.

Finally, whilst every attempt has been made to include all materials likely to be used in each application, absence of any material does not mean it can be assumed to be either safe or hazardous, nor to have either no or significant impact on the environment.

44

CHAPTER SIX

Application Sheets

Application Title

Roofs
1.1   Roof Coverings - Slates and Tiles
1.2   Roof Coverings - Sheets
1.3   Fascia, Soffit and Barge Boards
1.4   Verge - Pitched Roof
1.5   Roof Flashings
1.6   Rainwater Pipes and Gutters
1.7   Flat Roof Coverings

Insulation
2.1   Ground Floor Insulation
2.2   Cavity Wall Insulation
2.3   External Insulation
2.4   Internal Insulation Lining
2.5   Timber Framed Wall Insulation and Cavity Barriers
2.6   Pitched Roof Insulation
2.7   Flat Roof Insulation
2.8   Pipe Insulation
2.9   Hot and Cold Water Tank Insulation

Doors and Windows
3.1   Window/External Door Frames
3.2   Lintels and Arches
3.3   Cills and Thresholds
3.4   Glazing Fixing Systems
3.5   Sealants to Door and Window Frames
3.6   Fire Doors
3.7   External Doors
3.8   Internal Doors
3.9   Leaded Lights
3.10   Glazing Roof and Wall Structural Systems

Fittings and Finishings
4.1   Internal Non-Loadbearing Partitions
4.2   Ceiling and Wall Linings
4.3   Internal Wall Finishes
4.4   Architraves and Skirtings

Application Title

Roofs
4.5   Floorboarding
4.6   Floor Tile and Sheet
4.7   Carpet and Carpet Tiles
4.8   External Wall Finishes

Services
5.1   Above Ground Drainage
5.2   Underground Drainage
5.3   Hot and Cold Water Pipework
5.4   Cold Water Storage Tanks
5.5   Electrical Ducts and Conduits
5.6   Flues and Flue Pipes
5.7   Sanitary Fittings

External Works
6.1   External Pavings
6.2   Flat Roof Promenade Tiles
6.3   Boundary Fencing and Walling

Generic Materials
7.1   Concrete and Additives
7.2   Aggregates
7.3   Bricks and Blocks
7.4   Adhesives
7.5   Sealants
7.6   Timber
7.7   Timber Preservatives
7.8   Water Based Wall and Ceiling Paints
7.9   Trim Paints
7.10   Varnish, Wood Dyes, Wood Stains, Decorative Effects
7.11   Wall Paper and Wallpaper Pastes
7.12   Man Made Timber Boards
7.13   Damp Proof Membranes (DPM), Damp Proof Courses (DPC),
        Damp Proofing Systems
7.14   Building Membranes
7.15   Glass and Other Glazing Materials

Application 1.1
ROOF COVERINGS – SLATES & TILES

Typical Situation

slates laid double lap

battens

underfelting

rafters

Technical Requirements

Waterproof overlapping roof covering. Durable, impervious, rot and frost resistant. Must be resistant to ignition from fire exterior to the building, 30 year life absolute minimum, 60 plus preferred (relates to cost). Must either permit fixing holes to be made or come pre-drilled. Must not warp or bend. Must be compatible with cement mortars and metal flashings.

Decay and Degradation Factors

Normal weathering. Frost action. Fire. Abrasion from maintenance and foot traffic. Possible chemical attack by water run off from metal flashings.

Guidance Notes

The key alternatives are tiles or slates. The choice is determined primarily by aesthetic considerations, either from personal choice, or from local practice or planning requirements.

If slate appearance is required then technically, aesthetically and environmentally natural slate is to be preferred, especially if this is available from a U.K. source. Reuse of existing slate is technically sound and provides least environmental impact. For tiled roofs, concrete offers the best cost/durability/ environmental impact factor. As for slates, reused clay tiles offer the least environmental impact for a tiled roof. A market in salvaged concrete tiles does not exist due to relatively young age of this alternative and relatively low cost of new material.

Asbestos cement saw considerable use in the past, mainly as cheaper alternative to natural slate.

With existing roof coverings of asbestos cement, abrasion and rubbing down of the surface must be avoided. The risk of exposure to fibre from the material in-situ due to weathering is not thought to be sufficiently serious to recommend immediate removal. Careful removal and disposal is necessary at replacement and demolition and needs to be managed – see Chapter 7.

Alternative 6 is aimed primarily at short-life buildings where facsimile tile appearance is required.

Alternatives	Technical Comment	Rank
1 Natural Slate and slabs	Proven, very durable material. Non-combustible. Certain types, e.g. Yorkstone, very heavy requiring additional support.(*)	1 (2*)
2 Concrete tiles and slate (some referred to as interlocking tiles when laid single lap)	Interlocking concrete slates lighter than plain slates which require to be laid double lap (*). Both types give good durability and are non-combustible.	2 (3*)
3 Clay tiles	Pantiles – single lap. Plaintiles – double lap and heavier. Both types give good durability and are non-combustible.	2 3
4 Glass reinforced cement	Appearance similar to natural slate therefore has advantage as cheaper replacement for this material. Testing indicates durability will be comparable with asbestos cement. Non-combustible.	2
5 PVA cement slates	Appearance similar to natural slate therefore has advantage as cheaper replacement for this material. Testing indicates durability will be comparable with asbestos cement. Non-combustible.	2
6 Bitumen felt simulated slates	Appearance and durability considered to be much inferior to others. Combustible, fire properties vary. Otherwise see comments under applications 7.13 and 7.14.	8
7 Asbestos cement slate	No longer available.	-

Health Comment	Rank	Environmental Issues	Rank	Cost Rank
No significant risk foreseen to occupants.	0/0	Very good life expectancy, capable of reuse many times. Despolation by quarry waste. If slate or stone second hand, then upstream score 0.	2/0/0/0	332
No significant risk foreseen to occupants.	0/0	Lower manufacturing energy. Potential for recycling.	1/0/1/2	100
No significant risk foreseen to occupants.	0/0	Higher embodied energy and atmospheric emissions. A percentage can be reused. If tiles second hand, then upstream score 0 but lower durability anticipated.	2/0/1/1	251
Provided the diameters of the fibres have a low probability of being inhaled (see Chapter 2), subsequent abrasion of the product will present no risk to maintenance workers or to DIY occupants other than skin irritation. Machining generally produces high concentrations so protection and environmental control would be necessary.	0/0	Auto-claved products use more energy in process. Potential for recycling unclear.	2/0/1/2	293
The health status of PVA fibre is uncertain. If fibre with a high probability of being inhaled is released in weathering, cutting and attrition, maintenance workers and DIY occupants should restrict exposure to dust.	0/1	Some manufacturers' products have a proportion of quarry waste as aggregate. Polymers involve high energy processing. Potential for recycling unclear	2/0/1/2	N/A
No significant risk foreseen to occupants.	0/0	Shorter life than the alternatives.	2/0/2/2	102
Fibre release on ageing, cleaning, maintenance and disposal presents a potential risk to maintenance workers, DIY occupants, construction and demolition operatives.	1/3	Asbestos is a hazardous waste and needs to be disposed of properly. Asbestos in existing buildings needs to be identified, recorded, assessed for comparative risks and removed or perhaps treated or encased and managed as appropriate. The (-) classification refers to lack of availability/no longer used.	-/-/-/3	N/A

Application 1.2
ROOF COVERINGS - SHEETS

Typical Situation

ridge capping

sheeting

support rails

roof truss

side wall

Technical Requirements

Durable, impervious, rot, frost resistant material. Must permit site drilling for fixing and be compatible with fixing and sealants. Good fire properties, i.e. resistance to ignition, flame spread and penetration from external fire, all an advantage.

Decay and Degradation Factors

Normal weathering. Frost action. Fire. Abrasion from maintenance and foot traffic. Possible chemical attack by water run off from metal flashings and cement mortars.

Guidance Notes

These methods are rarely used for house roofs in the U.K., although technically feasible and see extensive use for housing in other countries. The most common use is of options 1, 2 and 7 for detached garages, particularly the pre-fabricated market, where the material may also be used for wall cladding.

Options 3 and 4 are marginally superior technically and 1 to 4 and 7 from an environmental perspective, although the concern over PVC may cause some to avoid options 3, 4 and 5.

Alternatives with good fire properties are to be preferred.

Guidance upon maintenance and the management of the disposal of existing asbestos cement sheeting provided on Application 1.1, Roof Coverings – Slates and Tiles, should be followed here and Chapter 7 provides further information.

Alternatives	Technical Comment	Rank
1 Calcium silicate with various non-asbestos fibres	Replacement for asbestos cement. Long-term durability claimed to be comparable. Similar performance. Good fire properties. Some varieties non-combustible.	3
2 Glass reinforced cement (GRC)	Replacement for asbestos cement. Long-term durability claimed to be comparable. Similar performance. Good fire properties. Some varieties non-combustible.	3
3 Plastic coated galvanized steel systems	Good durability expected from currently available sheeting. Slightly easier to fix than 1 and 2, using self drilling, self tapping fixings. Good fire properties. Advantage of single length sheet from ridge to eaves (max 6m).	2
4 Plastic coated aluminium systems	Good durability expected from currently available sheeting. Slightly easier to fix than 1 and 2, using self drilling, self tapping fixings. Good fire properties. Advantage of single length sheet from ridge to eaves (max 6m).	1
5 Polyvinyl-chloride (PVC)	Suitable for glazed areas. Ranking assumes use in this context. Variable fire properties.	3
6 Glass reinforced plastic (GRP)	Usual opaque material normally only used for special features. Rarely applied to domestic buildings except for glazed areas. Variable fire properties.	3
7 PVA cement	Recent development as replacement for asbestos cement. Long-term durability claimed to be comparable. Similar performance. Good fire properties. Some varieties non-combustible.	3
8 Asbestos cement	No longer used in new work, but available until recently for agricultural uses.	-

Health Comment	Rank	Environmental Issues	Rank	Cost Rank
No risk foreseen to occupants if undisturbed. Fibres may be released by abrasion, machining, maintenance or cleaning, or by the action of 'aggressive' water. Hazard to maintenance workers and DIY occupant depends on the nature and quantity of the fibre inhaled. See Chapter 2.	0/1	Moderate embodied energy. Dispolation from silicate quarrying and waste.	2/0/1/1	N/A
Provided the diameters of the fibres have a low probability of being inhaled, subsequent abrasion of the product, will present no risk to maintenance workers or to DIY occupants. Machining generally produces high concentrations and personal protection and environmental control would be necessary.	0/0	Moderate embodied energy. Pollution from cement manufacture.	2/0/1/1	N/A
No significant risk foreseen to occupants.	0/0	Marginally better life expectancy. Steel can be recycled. PVC coating.	2/0/1/1	385
No significant risk foreseen to occupants.	0/0	Marginally better life expectancy. High embodied energy. PVC coating.	2/0/1/1	100
No significant risk foreseen to occupants.	0/0	Problems from pollution during manufacture – but situation is improving. Problems at disposal. Shorter life expectancy.	2/0/1/2	285
Provided the diameters of the fibre have a low probability of being inhaled (see chap 2), subsequent abrasion of the product, will present little risk to maintenance workers or to DIY occupants. Machining generally produces high concentrations and personal protection and environmental control would be necessary.	0/0	Shorter life expectancy. Pollution from resin manufacture.	1/0/2/2	152
The health status of PVA is uncertain. If fibre with a high probability of being inhaled is released in weathering, cutting and attrition, maintenance workers and DIY occupants should restrict exposure to dust.	0/1	Polymers involve higher energy processing.	2/0/1/1	N/A
Fibre release on ageing, cleaning, maintenance and disposal presents a potential hazard to DIY occupants, maintenance, construction and demolition operatives.	1/3	Asbestos is a hazardous waste and needs to be disposed of properly. Asbestos in existing buildings needs to be identified, recorded, assessed for comparative risks and removed or perhaps treated or encased and managed as appropriate. The (-) classification refers to lack of availability/no longer used.	-/-/-/3	N/A

Application 1.3
FASCIA, SOFFIT AND BARGE BOARDS

Typical Situation

Tiles or slates

bargeboards

fascia

soffit *(under)*

gutter position

Technical Requirements

To close off space at the eaves between the edge of a roof and the top of an external wall. Span between and nail fix to rafters at normal spacing, 400mm – 600mm centres, without excessive shrinkage or sag. Allow provision of ventilation holes and permit decoration. Ease of cut and fit essential. Water resistance and good fire properties, i.e. not readily penetrated by or involved in fire an advantage.

Decay and Degradation Factors

Normal weathering. Abrasion from rubbing down prior to repainting. Wild life. Frost. Fire, particularly above window openings. Soffit is fairly well protected on underside of eaves.

Guidance Notes

From a technical standpoint options 1 and 2 are easier to cut and fix but require regular decoration to ensure durability. Non-combustible or low flame spread materials are preferred particularly for the soffit boards, especially adjacent to windows. Overhanging eaves give better protection to wall and adjacent openings.

Softwood from sustainable sources seems to offer least environmental impact, but this does not consider need for redecoration and should be judged alongside preservative or paint systems, see applications 7.8 and 7.9. The combined impact is less clear. Concern over PVC may cause some to avoid option 5.

Natural finish or painted asbestos board surfaces in existing buildings should not be abraded (sanded down), drilled or cut during alterations. Careful management of the removal and disposal is necessary on replacement and demolition – see Chapter 7.

Alternatives	Technical Comment	Rank
1 Plywood	Should be exterior WBP grade. Decoration required for protection, easy to cut and fit.	2
2 Softwood	Made up from single board or T & G boarding for wider soffit configuration. Decoration required for protection. Easy to cut and fit.	2
3 Calcium silicate board	Reasonable workability characteristics and fire properties. Some varieties are non-combustible. Board quality determines performance so designers should select accordingly.	1
4 PVA Cement Board	Good workability characteristics and fire properties. Some varieties are non-combustible. Board quality determines performance so designers should select accordingly.	1
5 Unplasticized polyvinyl-chloride (PVCu)	Some types use clip system, which requires soffit to be whole number of plank units in width as it is impracticable to cut planks along the length.	2
6 Glass Reinforced cement board	Good workability characteristics and fire properties. Some varieties are non-combustible. Board quality determines performance so designers should select accordingly.	1
7 Asbestos cement board	No longer used in new work but used in the past used for soffit boards.	-

Health Comment	Rank	Environmental Issues	Rank	Cost Rank
No significant risk foreseen to occupants.	0/0	The source of plywood is a key issue. Life expectancy related to maintenance – see paint. Disposal by combustion causes atmospheric pollution.	2/0/1/1	133+
No significant risk foreseen to occupants.	0/0	Life expectancy related to maintenance and preservative treatments – see application paint and wood preservatives 7.7 and 7.10. Disposal by combustion causes atmospheric pollution.	1/0/1/0	100+
Mineral fibres may be released by abrasion, machining, maintenance or cleaning, or by aggressive water. The order of hazard for the maintenance worker and DIY occupant depends on the nature and quantity of the fibre inhaled. See Chapter 2.	0/1	Moderate embodied energy. Dispoilation from silicates quarrying and waste.	2/0/1/1	247++*
The health status of PVA fibre is uncertain. If fibre with a high probability of being inhaled is released in weathering, cutting and attrition, maintenance workers and DIY occupants should restrict exposure to dust.	0/1	Polymers involve high energy processing.	2/0/1/1	130++*
No significant risk foreseen to occupants.	0/0	Problems from pollution during manufacture – but situation is improving. Problems at disposal.	2/0/1/2	185++*
Provided the diameters of the fibres have a low probability of being inhaled (see chap. 2), subsequent abrasion, or machining of the product, will present little or no hazard to maintenance workers or to DIY occupants other than skin irritation. Machining generally produces high concentrations and personal protection and environmental control would be necessary.	0/0	Moderate embodied energy. Impact from cement manufacture.	2/0/1/1	N/A
Fibre release on ageing, cleaning, maintenance and disposal present a potential risk to maintenance workers, DIY occupants, construction and demolition operatives.	1/3	Asbestos is a hazardous waste and needs to be disposed of properly. Asbestos in existing buildings needs to be identified, recorded, assessed for comparative risks and removed or perhaps treated or encased and managed as appropriate. The (-) classification refers to lack of availability/no longer used.	-/-/-/3	N/A

+ 150mm wide
++ 1000mm wide
* supply only

Application 1.4
VERGE – PITCHED ROOF

Typical Situation

batten
tiles/slates
cement motar pointing
rafter
verge undercloak –bedded on mortar
gable wall

Technical Requirements

Board material or special tiles to provide neat joint at junction of roof and gable wall. Water and rot-proof, frost resistant material required. Overhanging verge detail requires nailability for fixing. Flush verge detail shown requires key/compatibility with cement mortar.

Decay and Degradation Factors

Flush verge: Normal weathering. Frost action. Abrasion from raking out mortar.
Overhanging verge: Fairly protected on underside of verge. Abrasion from rubbing down prior to repainting. Wild life. Fire.

Guidance Notes

The key alternatives are overhanging and flush verge construction. Overhanging verge construction offers better protection to the wall and adjacent openings and the nature of the detailing influences the material choice. For overhanging verge construction the detail is similar to eaves construction and therefore comments on Application 1.3 apply. With the flush eaves detail shown all alternatives other than 1 and 2 are suitable from a technical viewpoint.

Designers should assure themselves regarding the frost resistance of alternatives 3 and 4 as manufacturers produce a variety of types. For flush verges option 8 has become popular to match finish with various roof tile systems.

As the quantity of material required for each house is small, the minor differences in environmental impact are of little significance

Existing examples of asbestos cement in this application may have been decorated. Abrasion (rubbing down) prior to decoration should be avoided. The most practical way or removing mosses, lichens and mould growth is by chemical cleaners.

This material should be disposed of properly and needs to be properly managed and disposed of in accordance with waste disposal regulations – see Chapter 7.

Alternatives	Technical Comment	Rank
1 Plywood	Only suitable for overhanging verge details where protected from weather. Must be decorated for protection. Easy to cut and fix.	2
2 Softwood	Made up from single board or T & G boarding for wider soffit configuration. Decoration required for protection. Easy to cut and fix. Only suitable for overhanging verge details where protected from weather. Must be decorated for protection.	2
3 Calcium silicate board	Manufacturers now produce purpose designed boards for external application which offer comparable durability with asbestos cement. Good workability characteristics.	1
4 PVA cement board	Manufacturers now produce purpose designed boards for external application which offer comparable durability with asbestos cement. Good workability characteristics.	1
5 Glass reinforced cement board	Good workability characteristics and fire properties. Some varieties are non-combustible. Board quality determines performance so designers should select accordingly.	1
6 Unplasticized polyvinyl-chloride (PVCu)	Made up from clip fix interlocking sections. Has advantage of eliminating need for mortar pointing. Long-term durability expected to be satisfactory. Proprietary system only suitable for use with compatible tile profiles.	1
7 Natural slate	Very durable material compatible with mortar. Disadvantage of limited length of slates requiring overlaps and careful workmanship.	1
8 Tiles (clay or concrete)	Similar durability to roof covering. Replacement awkward when damaged. Disadvantage of limited length of tiles requiring overlaps and careful workmanship.	1
9 Asbestos cement	No longer used in new work.	1

Health Comment	Rank	Environmental Issues	Rank	Cost Rank
No significant risk foreseen to occupants.	0/0	The source of plywood is a key issue. Life expectancy related to maintenance – see application paint 7.9 and 7.10. Disposal by combustion causes atmospheric pollution.	2/0/1/1	378
No significant risk foreseen to occupants.	0/0	Life expectancy related to maintenance and preservative treatments – see application paint and wood preservatives 7.7, 7.9 and 7.10. Disposal by combustion causes atmospheric pollution.	1/0/1/0	413
No risk foreseen if undisburbed. Fibres may be released by abrasion, machining, maintenance or cleaning, or by the action of 'aggressive' water. Hazard to maintenance and DIY occupant depends on the nature and quantity of fibre inhaled.	0/1	Moderate embodied energy. Silicates from quarrying.	2/0/1/1	N/A
The health status of PVA fibre is uncertain. If fibre with a high probability of being inhaled is released on weathering, cutting and attrition, maintenance workers and DIY occupants should restrict exposure to dust.	0/1	Polymers involve high energy processing.	2/0/1/1	667
Provided the diameters of the fibres have a low probability of being inhaled (see chap. 2), subsequent abrasion, or machining of the product, will present little risk to maintenance workers and DIY occupants other than skin irritation. Machining generally produces high concentrations and personal protection and environmental control would be necessary.	0/0	Moderate embodied energy. Pollution from cement manufacture.	2/0/1/1	N/A
No significant risk foreseen to occupants.	0/0	Problems from pollution during manufacture – situation is improving. Problems at disposal.	2/0/1/2	100
No significant risk foreseen to occupants.	0/0	Good life expectancy. Despolation by quarry waste.	2/0/0/2	268
No significant risk foreseen to occupants.	0/0	Clay – high embodied energy and atmospheric emissions. Concrete – lower embodied energy.	2/0/1/2	107
Fibre release on ageing, cleaning, maintenance and disposal presents a potential hazard to maintenance workers, DIY occupants, construction and demolition operatives.	1/3	Asbestos is a hazardous waste and needs to be disposed of properly. Asbestos in existing buildings needs to be identified, recorded, assessed for comparative risks and removed or perhaps treated or encased and managed as appropriate. The (-) classification refers to lack of availability/no longer used.	-/-/-/3	N/A

Application 1.5 ROOF FLASHINGS	Alternatives	Technical Comment	Rank
Typical Situation 	1 Lead	In many ways the ideal material combining good malleability and workability with excellent durability when properly applied.	1
	2 Copper	Generally slightly less malleable than lead giving rise to possible problems with complex shapes. Do not mix with other metals. Detailing requires care to avoid staining caused by rainwater run-off.	2
	3 Zinc	Generally slightly less malleable than lead giving rise to possible problems with complex shapes. Avoid contact with copper.	2
	4 Aluminium	Malleable grades of aluminium alloy now available. Care needed in selection of quality of aluminium as it can be attacked by water run-off from cementitious material. Electrolytic action possible in polluted atmospheres, particularly when in contact with ferrous metals.	3
	5 Stainless steel	Rarely used in the past for flashings. Malleable grades now available but there may still be problems with complex shapes.	2
	6 Bitumen reinforced with aluminium foil	Available with self-adhesive backing and intended as short term repair methods. Useful in this context. Combustible.	8

Technical Requirements

Impervious material to close off gaps at junctions of roof covering and other features. Durability must equal that of roof covering. Malleability a big advantage but non non-malleability can be offset to some extent by the availability of a range of preformed sections. Must accommodate movements in the roof. Non-combustibility an advantage. Flashings should not detract from the appearance of the main roof covering or cause unsightly staining.

Decay and Degradation Factors

Normal weathering. Wild life. Frost. Bi-metallic corrosion can occur with dissimilar metals in contact but normally avoided by good building practice.

Certain grades of aluminium can be attacked by run-off from cementitious materials.

Guidance Notes

Lead finds extensive usage for roof flashings, due to its combination of excellent durability with malleability allowing it to be easily worked to form complex profiles; however, other metals are available which can be used as an alternative in most circumstances. None of the materials in flashing applications pose a significant health hazard to the occupier. The main concern is over careless disposal of lead. However, the environmental scores for this and other metal options assume that the material is recycled at end of life and not disposed in landfill.

Zinc is the most commonly available option other than lead but alternatives 2 and 4 may also be satisfactory depending upon the exact detail of the flashing application. Complex pantile profiles may still pose a problem with these materials. The main criticism of copper is the risk with certain lighter coloured open textured tiles and slates of unsightly staining due to water run-off from the flashing. This is obviously not a problem with most valley or internal gutter positions. Stainless steel offers the lowest environmental impact but at higher cost.

It is important that occupants do not collect run-off from lead covered roofs and roofs with lead flashings for drinking purposes or for watering vegetables. Where this is expected, extra consideration should be given to the position of the flashing on the roof as there is likely to be a more concentrated flow over valleys and parapet gutters than other locations. Other alternatives might be used in this location.

Appearance may be a critical factor in selection as the majority of the options have their own distinctive character which may be used to blend or contrast with the general roof covering.

Health Comment	Rank	Environmental Issues	Rank	Cost Rank
Provided run off water is kept away from the occupants and their garden produce, no significant risk foreseen to occupants.	0/0	High embodied energy. Pollution from manufacturing. Score assumes 100% recycled content.	2/0/0/0	127
No significant risk foreseen to occupants.	0/0	High embodied energy. Pollution from manufacturing. Score assumes 100% recycled content.	2/0/0/0	122
No significant risk foreseen to occupants.	0/0	As above but there is a question over recycling percentage. Lower embodied energy than other metals from primary production.	2/0/0/0	113
No significant risk foreseen to occupants.	0/0	High embodied energy. Pollution from manufacturing. Score assumes 100% recycled content. Note durability risks/see technical comments.	2/0/0/0	100
No significant risk foreseen to occupants.	0/0	Score assumes UK source with 100% recycled content. Material from other sources would rank as other metals alternatives (options 1 to 4).	1/0/0/0	201
Although all coal tar products are carcinogenic, natural bitumen is less so, and exposures will be extremely low. No significant risk foreseen to occupants.	0/0	Poor durability. Potential of pollution at disposal.	1/0/2/2	N/A

Application 1.6 RAINWATER PIPES AND GUTTERS	Alternatives	Technical Comment	Rank
Typical Situation tiles/slates gutter roof fascia elbows/joints brackets	1 Cast iron	Durable but brittle material. Will support loads from ladder. Needs regular painting to maintain good appearance.	2
	2 Unplasticized polyvinyl-chloride (PVCu)	Lightweight easily worked material. Wide range of proprietary systems now available. Improved products now offer adequate durability. Painting not required.	1
	3 Aluminium	Durable lightweight material. Normally available only through specialist fixers. Seamless pattern. Painting not required.	1
Technical Requirements Durable and impervious gutter and downpipe. Must span recommended bracket spacing, provide easy jointing and cutting and permit decoration if required. Jointing system must cope with thermal movements. Ability to support ladder during maintenance an advantage as is resistance to impact at base of downpipe. Must be compatible with most building materials.	4 Timber (gutters only)	Regular painting required to ensure durability. Timber needs to be vacuum impregnated with preservative for long term durability. Will support loads from ladder.	3
Decay and Degradation Factors Normal weathering. Damage from impact near ground level. Attack by weak acid solutions in polluted environments. Mould growth, vegetation and wild life. Plastics materials gradually degraded by U.V. radiation.	5 PVA cement (asbestos replacement)	Needs regular painting to give good appearance. Lighter weight than cast iron. Durability comparable with asbestos cement, impact resistance slightly improved.	3
Guidance Notes Selection is based mainly upon appearance and cost. PVCu now ubiquitous due to good appearance, cost factors and adequate technical performance in the vast majority of domestic applications. Environmental concerns over pvc. may cause some to avoid this material. In this context cast iron and aluminium offer some advantages. Cast iron and timber gutters are usually only used for replacement on older architectural merit where continuity of appearance is required. Use of lead for caulking spigot and socket joints in cast iron work is not recommended on health grounds. Purpose made seamless aluminium can be produced to match cast iron gutter profiles. Existing asbestos cement material must be disposed of properly and needs to be properly managed and disposed of in accordance with waste disposal regulations - see Chapter 7.	6 Asbestos cement	No longer used in new work. Had limited application in house building.	3

Health Comment	Rank	Environmental Issues	Rank	Cost Rank
No significant risk foreseen to occupants. (Exposure to lead based primer and lead caulking in old structures in maintenance and DIY occupants would present a small risk.)	0/0	Raw material source is critical, eg. pig iron is smelted with charcoal in some parts of Brazil. See technical comment on maintenance for good durability.	2/0/0/0	320*
No significant risk foreseen to occupants.	0/0	Problems with pollution from manufacture but situation is improving. Problems at disposal.	2/0/1/2	129*
No significant risk is foreseen to occupants.	0/0	High embodied energy. Pollution from manufacturing. High recycled content.	2/0/0/0	277*
No significant risk is foreseen to occupants.	0/0	Life expectancy related to maintenance and preservative treatments – see applications paints and wood preservatives 7.7, 7.9 and 7.10. Disposal by combustion causes atmospheric pollution.	1/0/1/0	100+
The status of PVA fibre is uncertain. Weathering, cutting and attrition will release fibre. If maintenance workers and DIY occupants are exposed to inhalable fibre, a small health risk will result (see chap. 2).	0/1	Polymers involve high energy processing.	2/0/1/1	N/A
Fibre release on ageing and maintenance in old material will expose maintenance workers and DIY occupants, construction and demolition operatives to an inhalation risk.	1/3	Asbestos is a hazardous waste and needs to be disposed of properly. Asbestos in existing buildings needs to be identified, recorded, assessed for comparative risks and removed or perhaps treated or encased and managed as appropriate. The (-) classification refers to lack of availability/no longer used.	-/-/-/3	N/A

+ gutter only
* averaged

Application 1.7 FLAT ROOF COVERINGS	**Alternatives**	**Technical Comment**	**Rank**
Typical Situation vapour barrier set in bitumen compound rigid insulation asphalt, 20mm in two layers or 3 layer felt in bitumen compound stone chippings set in bitumen compound structural deck WARM ROOF	1 Bitumen Felt Systems	3 layer systems	
	a) Textile fibre base (jute)	Very poor durability, not used for roofs of good quality.	8
	b) Glass fibre base	Glass fibre base is rot proof giving improved durability. Not suitable for mechanical anchorage to deck.	3
	c) Polyester fibre base	Preferred bitumen/fibre based system.	2
Technical Requirements Waterproof flexible sheet materials layed in various combinations set in bitumen compound. Durable, impervious, rot and frost-resistant material required. Roof covering system must be resistant to ignition from fire exterior to the building. Single layer also used as underfelting on pitched roofs, normally textile fibre based. All applications must resist thermal movements of daily and seasonal cycles. Must be compatible with cement mortars and metal flashings.	2 High Performance Felt Systems		
	a) Pitch polymer	Good performance feasible with 2 layer systems.	1
	b) Polyester base	Good performance feasible with 2 layer systems.	1
Decay and Degradation Factors Normal weathering – note protected from UV by solar protective layer (paint, white spar chippings or insulation). Frost action. Possible chemical attack by run off from metal flashings. Abrasion from foot traffic and maintenance. Mould growth, vegetation and wild life. Fire.	3 Single Layer Systems		
	a) Polyvinyl-chloride (PVC)	Performance related to workmanship and quality of jointing.	2
Guidance Notes It is extremely difficult to give detailed guidance as roof covering specifications must be related to the type of decking, type of insulation and its position within the roof structure.	b) Polyurethane (PU)	Performance related to workman-ship and quality of jointing.	2
High performance systems offer the best performance of the bitumen based alternatives. Single layer systems can perform well provided workmanship is of high quality. Asphalt performs well when properly isolated from roof movements. Turfed roof is an option utilising one of these alternatives as the waterproofing layer. Turf improves durability by isolating membranes from the U.V. degradation and temperature fluctuations. Additional structural support required.	c) Butyl Rubber	Performance related to workman-ship and quality of jointing.	2
From an environmental perspective the bitumen based systems are to be preferred.	4 Asphalt	Normally used on decks of heavyweight construction such as concrete roofs. Problems can arise with roofs of lightweight construction due to excessive movement.*	1
			*3
The main problem with existing roofs is identifying whether asbestos fibre based felt has been used. Use of this material was discontinued in the 1980's and it is believed that these felts never accounted for more than 20% of the roofing felt utilized in the UK. Design drawings and specifications may be a source of information otherwise investigation may be necessary. Avoid unnecessary abrasion upon repair, maintenance and removal. Asbestos fibre material must be disposed of carefully and needs to be managed properly and disposed of in accordance with waste regulations – see Chapter 7.	5 Standing seam troughed steel and aluminium cladding	Not a true alternative for the flat roof systems but can now be laid to a 4° pitch provided joints properly sealed. Noise drumming may be a problem.	1

Health Comment	Rank	Environmental Issues	Rank	Cost Rank
Although all coal tar products are carcinogenic, natural bitumen is less so, and exposures during installation will be extremely low. No significant risk to occupants from bitumens. Minor risk during DIY maintenance and repair.		The main environmental impacts accrue from bitumen. Manufacturing pollution and embodied energy. For option 1a – see technical comment on durability.		
a) No significant risk to occupants is foreseen.	0/0		1/0/2/1	N/A
b) Only inhaled fibres will present a potential risk (see chap. 2). Weathering and attrition will release fibre, but exposure will be low. No significant risk to occupants foreseen.	0/1		1/0/1/1	100
c) Provided fibre not inhalable (see chap. 2) no significant risk to occupants foreseen.	0/1		1/0/1/1	118
a) No significant risk to occupants foreseen.	0/0	Manufacturing pollution and embodied energy.	1/0/1/1	169
b) No significant risk to occupants foreseen.	0/0		1/0/1/1	155
a) No significant risk to occupants foreseen.	0/0	Pollution from manufacture. Problems from disposal.	2/0/1/1	155
b) No significant risk to occupants foreseen.	0/0	Pollution from manufacture. Problems from disposal.	2/0/1/1	149
c) No significant risk to occupants foreseen.	0/0	Grading assumes synthetic rubber.	2/0/0/0	178
Although all coal tar products are carcinogenic, natural bitumen is less so. No significant risk to occupants foreseen.	0/0	Main environmental impacts from bitumen. Manufacturing pollution and embodied energy. Durability can be excellent – see technical comments.	2/1/0/1	169
No significant risk to occupants foreseen.	0/0	Steel: Pollution from manufacture. Recycled content is a consideration.	1/0/0/0	+
		Aluminium: High-embodied energy. Pollution from manufacturing. Normally high recycled content.	2/0/0/0	+

+ refer to specialist supplier

Application 2.1 GROUND FLOOR INSULATION	Alternatives	Technical Comment	Rank
Typical Situation concrete — UNDER SLAB INSULATION floor slab — rigid insulation hardcore + sand blinding — DPM OVER SLAB INSULATION — chipboard — rigid insulation floor slab hardcore + sand blinding — DPM SUSPENDED FLOOR INSULATION — floor boards floor joists — quilt insulation supported on galvanised metal mesh	1 Polyurethane foam board	Combustible. Reinforce screeds to resist cracking due to board movement.	1
	2 Polyisocy-anurate foam board	Ditto. Slightly better fire properties than other foamed plastic except phenolic.	1
	3 Extruded polystyrene foam board	Combustible. Reinforce screeds to resist cracking due to board movement.	1
	4 Expanded polystyrene beadboard	Combustible. Reinforce screeds to resist cracking due to board movement.	2
Technical Requirements Rot resistant insulation material capable of withstanding floor loads (except option 8). Water resistance and non-combustibility an advantage. Must be dimensionally stable.	5 Phenolic foam	Best fire properties of plastic foams. Reinforce screeds to resist cracking due to board movement.	1
Decay and Degradation Factors Vermin, intersitial condensation, fire. Chemical reaction with pvc sheathing to electric wiring. Plumbing and electrical alterations. Flooding in low lying areas.	6 Corkboard	Combustible. Reinforce screeds to resist cracking due to board movement.	4
Guidance Notes Material options are related to constructional method. Boarded material is generally used under slabs or below screeds. Quilt or non-combustible board can be used for suspended timber ground floors but needs to be fixed in place, e.g retained with mesh between joints.	7 Woodfibre board (softboard)	Combustible. Reinforce screeds to resist cracking due to board movement.	7
Overall thermal performance is a function of the K value of the material and detail at edges and other features to eliminate cold bridges. Laminates of chipboard and plastic foams are available to lay over slabs, but this adds to the potential fire load of the interior.	8 Mineral and glass fibre quilt	Non-combustible (suspended timber floor only).	2
The higher upstream score for the plastic based material is primarily due to high embodied energy but, with housing, where buildings are normally long-lived it is important to note that the environmental impact of insulation will be relatively minor when considered against the reduction in impact that the material contributes to life-time energy savings. Therefore, designers should seek first to maximise thermal performance in the building.	9 Mineral and glass fibre board	Non-combustible. Reinforce screeds to resist cracking due to board movement.	2
From an environmental perspective corkboard and woodfibre board offer the least impact but need to be kept dry. M.M.M.F. is marginally preferable to plastic foams. Extruded foam boards and cellular glass offer superior water resistance and should be used in exposed situations below ground. The use of materials that liberate volatile compounds require either an impermeable floor above or for the area to be well ventilated. In the case of materials containing inhalable mineral fibres, an impermeable floor is indicated. Care must be taken by electricians and plumbers who subsequently need to enter the under floor space.	10 Cellular glass	Non-combustible. Reinforce screeds to resist cracking due to board movement.	1

Health Comment	Rank	Environmental Issues	Rank	Cost Rank
Intact fully reacted polymers present no significant health risk to occupants. Where ventilation is poor and volatile compounds are released, concentrations will build up and may affect occupants, thereby elevating the ranking to 0/2.	0/0	High embodied energy. Pollution from manufacture and eventual disposal. Older material in existing buildings is likely to contain CFCs.	2/0/0/2	784
Ditto	0/0	Ditto	2/0/0/2	876
Ditto	0/0	Ditto	2/0/0/2	100
Ditto	0/0	High embodied energy. Pollution from manufacture and eventual disposal. CFCs have never been used in manufacture of this product.	2/0/0/2	133
Ditto	0/0	High embodied energy. Pollution from manufacture and eventual disposal. Older material in existing buildings is likely to contain CFCs.	2/0/0/2	631
No significant risk foreseen to occupants.	0/0	Low embodied energy. Durability assumes application in dry conditions. Pollution if disposed of by combustion.	0/0/0/1	694
No significant risk foreseen to occupants.	0/0	Low embodied energy. Chemical additives may be used to resist moisture. Pollution if disposed of by combustion.	0/0/1/1	483
So long as fibres are contained, no significant risk to occupants is foreseen. In the event of inhalable fibre being liberated into occupied areas as a result of maintenance or repair, or by building vibration and air currents, there is a potential health risk (see chap. 2).	0/1	Resin binders. Pollution during manufacture and disposal.	1/0/0/2	100
Ditto	0/1	Ditto	1/0/0/2	721
No significant risk to occupants.	0/0	High embodied energy. Pollution from manufacture. Potential for recycling depending on method of fixing.	2/0/0/1	N/A

Application 2.2
CAVITY WALL INSULATION

Typical Situation

built-in or injected post construction

25mm insulation

75mm insulation

air gap

built-in only

FULL FILL   PARTIAL FILL

Technical Requirements

Rot resistant insulating material which combined with other wall materials must give a minimum U value to comply with the latest Building Regulations. Material to be water resistant, not encourage moisture penetration. Compatibility with masonry materials and mortar essential. Must not cause premature corrosion of wall ties. Non-combustibility an advantage.

Decay and Degradation Factors

Frost, fire, water and vermin. Electrical, plumbing and general building alterations can cause access to insulation. Gas or other vapours can be released into habitable rooms. Possible vapour release from insitu foam types.

Guidance Notes

In terms of weather resistance the decision on insulation is based upon the expected exposure (i.e. rainfall, location, local topography) together with the materials comprising each leaf of the wall. In cases of normal or sheltered exposure both partial and full fill are suitable provided workmanship is adequate. In cases of extreme exposure partial fill retaining 50mm minimum cavity is to be preferred in principle.

In terms of thermal performance the choice will be influenced by the thermal resistant of the entire wall and the relationship between capital and running costs. However, for many projects limited, due to capital cost considerations, to the minimum standards of the building regulations there is little to choose between the alternatives. The upstream score for the plastic based materials represents a higher 'upstream' impact primarily due to high embodied energy but, with housing, where buildings are normally long-lived it is important to note that the environmental impact of insulation will be relatively minor when considered against the reduction in impact that the material contributes to life-time energy savings. Therefore, designers should seek first to maximise thermal performance in the building. However, from an environmental perspective, increased thickness is essential. Thicknesses of the order of 200mm are justifiable and quite feasible technically. Similar U values can be achieved with less thickness using some of the plastic insulants. At this level of performance reductions in life-time energy consumption considerably outweigh differences in environmental impacts from the material alternatives.

Poor workmanship can be a significant cause of damp penetration with both fill configurations (as it can be with walls without added insulation). Boarded insulation requires particularly good workmanship by the bricklayer. Poor sealing around joist ends, at eaves and around openings, etc, may also permit loose fibre, foam and fume ingress to the interior with subsequent health implications.

In general terms, non-combustible material has an advantage. For retro fit, blown cellulose fibre, alternative B5, appears therefore to offer the best cost/technical/health/environmental/buildability compromise provided workmanship is adequate. It should be noted that non-combustibility is not mandatory and in this location the protection offered by the inner masonry leaf is considered to reduce the risk from toxic fume to an extremely low level.

Plastic wall ties are advantageous in reducing cold bridging. Proper detailing around openings and at eaves is also necessary to reduce cold bridging.

Alternatives	Technical Comment	Rank
A Built-in as Wall Proceeds		
1 Mineral wool fibre board	Non-combustible.	2
2 Expanded polystyrene bead-board	Combustible, but protected by wall seal cavity at eaves and around openings. Available with various fire properties	2
3 Extruded polystyrene board	Waterproof material. Combustible, but protected by wall seal cavity at eaves and around openings. Available with various fire properties.	2
4 Polyurethane foam board	Ditto	2
5 Polyisocy- anurate board	Ditto. Slightly better fire properties than other foamed plastics except phenolic.	2
6 Phenolic foam	Water-proof material. Better fire properties than other foamed plastics.	2
B Injected after Construction		
1 Urea- formaldehyde foam	Combustible, but protected by wall. Poor application leads to cracks in foam, which in turn can cause water penetration. Needs careful workmanship to seal inner leaf to reduce off gassing to interior. Site quality control vital.	3
2 Polyurethene foam	Primarily used to remedy wall tie failure as well as to insulate.	1
3 Bonded polystyrene beads	Adhesive injected with beads overcomes problem of loose fill. Combustible, but protected by wall.	1
4 Blown mineral fibre	Non combustible.	1
5 Blown cellulose fibre	Protection from fire, insects and fungi given by Boron.	1

Health Comment	Rank	Environmental Issues	Rank	Cost Rank
No significant risk to occupants is foreseen while all inhalable fibres are contained. In the event of inhalable fibres being liberated into occupied areas, especially during maintenance, this risk may be elevated to 0/2.	0/0	Resin binders. Pollution during manufacture and disposal.	1/0/0/2	124
No significant risk to occupants is foreseen.	0/0	High embodied energy. Pollution from manufacture and eventual disposal. Does not contain CFCs.	2/0/0/1	128
No significant risk to occupants is foreseen.	0/0	High embodied energy. Pollution from manufacture and eventual disposal. Older material in existing buildings is likely to contain CFCs.	2/0/0/2	100
No significant risk to occupants is foreseen.	0/0	High embodied energy. Pollution from manufacture and eventual disposal. Older material in existing buildings is likely to contain CFCs.	2/0/0/2	664
No significant risk to occupants is foreseen.	0/0	High embodied energy. Pollution from manufacture and eventual disposal. Older material in existing buildings is likely to contain CFCs.	2/0/0/2	739
No significant risk to occupants is foreseen.	0/0	High embodied energy. Pollution from manufacture and eventual disposal. Older material in existing buildings is likely to contain CFCs.	2/0/0/2	537
Exposure to formaldehyde is associated with acute eye, nose, throat complaints, and it is also considered to be a human carcinogen. Defects of installation technique, permeability or breaches in the internal wall, and reduced general ventilation, will allow irritant levels to build up in newly insulated buildings. Installation and design will determine whether the appropriate rank is 0/0 or 3/3.*	See Health Comment	High embodied energy. Pollution from manufacture and eventual disposal.	2/1/0/2	176
No significant risk to occupants is foreseen.	0/0	High embodied energy. Pollution from manufacture and eventual disposal. Older material in existing buildings is likely to contain CFCs.	2/0/0/2	176
No significant risk to occupants is foreseen.	0/0	High embodied energy. Pollution from manufacture and eventual disposal. Does not contain CFCs.	2/0/0/2	N/A
No significant risk to occupants is foreseen while all inhalable fibres are contained. In the event of inhalable fibres being liberated into occupied areas, especially during maintenance, this risk may be elevated 0/2.	0/0	Resin binders. Pollution during manufacture and disposal.	1/0/0/2	132
No health hazard foreseen for occupants so long as bulk material is contained. Too little is known of the material to be sanguine. If there is a possibility of exposure to substantial amounts of inhalable cellulose fibre, the risk may be elevated to 0/2.	0/0	Slight risk from Boron. Low embodied energy.	0/0/0/0	N/A

Application 2.3
EXTERNAL INSULATION

Typical Situation

insulation

mechanical fixing

cement/sand render reinforced with galvanised mesh or glass fibre

wall

VERTICAL SECTION

Technical Requirements

Insulation applied externally to improve performance of existing structures. Ability to take mechanical fixing or adhesive fixing depending on application. Rot-proof, with adequate impact strength. Finishes usually supported on stainless steel fixings and mesh for renderings.

Decay and Degradation Factors

Frost, wild life, vandalism, and accidental damage. Cracks in finish caused by thermal movement. Damp penetration through finish. Fire.

Guidance Notes

Most systems require render finish for appearance, fire or weather protection. Thus, the key technical selection factors are the resistance to moisture, and fire. Cellular glass is the ideal, but is very expensive. Beyond this there is little to choose between alternatives both technically and environmentally. M.M.M.F. has a slight environmental advantage.

The upstream score for the plastic based materials represents a higher embodied energy but, with housing, where buildings are normally long-lived it is important to note that the environmental impact of insulation will be relatively minor when considered against the reduction in impact that the material contributes to life-time energy savings. Therefore, designers should seek first to maximise thermal performance in the building.

Options 1-5 may yield toxic fumes, but by nature of the position in this application occupants are unlikely to be exposed.

Alternatives	Technical Comment	Rank
1 Polyurethane foam board	Combustible – may be protected by rendering or other class O cladding*. High board movement – employ stainless steel carrier mesh or GRC rendering to resist cracking.	2
2 Extruded polystyrene foam board	Combustible – may be protected by rendering or other class O cladding*. High board movement – employ stainless steel carrier mesh or GRC rendering to resist cracking.	2
3 Polyisocy-anurate foam board	Combustible – slightly better fire properties than the other foamed plastics except phenolic. May be protected by rendering or other class O cladding*. High board movement – employ stainless steel carrier mesh or GRC rendering to resist cracking.	2
4 Phenolic foam board	Slightly better fire properties than other foamed plastics – may be protected by rendering or other class O cladding*. High board movement – employ stainless steel carrier mesh or GRC rendering to resist cracking.	2
5 Expanded polystyrene beadboard	Combustible – may be protected by rendering or other class O cladding*. High board movement – employ stainless steel carrier mesh or GRC rendering to resist cracking.	2
6 Mineral and glass fibre board	Non-combustible.	2
7 Cellular glass	Non-combustible.	1
	* Class O protection is required when building is within 1m of relevant boundary or if over 15m high.	

Health Comment	Rank	Environmental Issues	Rank	Cost Rank
No significant risk to occupants foreseen.	0/0	High embodied energy. Pollution from manufacture and eventual disposal. Older material in existing buildings contains CFCs.	2/0/0/2	784
No significant risk to occupants foreseen.	0/0	High embodied energy. Pollution from manufacture and eventual disposal. Older material in existing buildings contains CFCs.	2/0/0/2	100
No significant risk to occupants foreseen.	0/0	High embodied energy. Pollution from manufacture and eventual disposal. Older material in existing buildings contains CFCs.	2/0/0/2	876
No significant risk to occupants foreseen.	0/0	High embodied energy. Pollution from manufacture and eventual disposal. Older material in existing buildings contains CFCs.	2/0/0/2	136
No significant risk to occupants foreseen.	0/0	High embodied energy. Pollution from manufacture. Does not contain CFCs.	2/0/0/2	133
No significant risk to occupants foreseen. (In the event of inhalable fibres being released into the breathing zone of maintenance workers or DIY activists, an elevation of risk may result (see chap. 3).)	0/0	Resin binders. Pollution during manufacture and disposal.	1/0/0/2	721
No hazard to occupants.	0/0	High embodied energy. Pollution from manufacture. Potential for recycling dependent on method of fixing.	2/0/0/1	N/A

Application 2.4 INTERNAL INSULATION LINING	Alternatives	Technical Comment	Rank
Typical Situation BATTEN FIXING — 50×50 GW. battens mechanically fixed — wall — insulation — plasterboard "DAB" FIXING — wall — adhesive dabs — insulation — support fixing nails — plasterboard laminate PLAN SECTION	1 Polyurethane foam board (PBL*)	Combustible – should be protected by plasterboard or class O lining.	2
	2 Polyisocyanurate foam board (PBL*)	Combustible – slightly better fire properties than other foamed plastics except phenolic. Should be protected by plasterboard or class O lining.	2
Technical Requirements Insulation to improve the insulation of existing structures or as part of the insulation provided in new construction. Rot proof, water resistance and non-combustibility an advantage.	3 Phenolic foam board (PBL*)	Combustible, but burns with difficulty, little smoke and toxic fume and lower flame spread. Best fire properties of foamed plastics.	2
Decay and Degradation Factors Vermin, dampness and interstitial condensation. Normal wear and tear. Fire. Chemical reaction with PVC wiring. Plumbing and electrical alterations.	4 Expanded polystyrene beadboard (PBL*)	Combustible – should be protected by plasterboard or class O lining.	2
Guidance Notes These materials provide a means to retrofit insulation to existing buildings without damage to the external appearance. The insulation receives some protection from the plasterboard lining in the event of a fire so it is desirable to use a non-combustible insulation in this application. Therefore mineral and glass wool seem more suitable, although plasterboard/foam laminates have a slight buildability advantage.	5 Mineral and glass wool quilt	Non-combustible. Quilts easy to install without cutting and fitting between fixing battens.	1
There are health concerns over off-gassing of volatile agents in alternatives 1-3 and fibre release in 5 and 6. Good workmanship is essential with this type of construction to ensure effective installation of the vapour barrier and thus eliminate potential interstitial condensation problems. The vapour barrier should eliminate fibre ingress to the interior and delay transmission of volatile gases.	6 Mineral and glass fibre batts.	Non combustible.	1
From the health perspective option 4 has a slight advantage whereas in environmental terms M.M.M.F. options 5 and 6 have an advantage. The upstream score for the plastic based materials represents a higher embodied energy but, with housing, where buildings are normally long-lived it is important to note that the environmental impact of insulation will be relatively minor when considered against the reduction in impact that the material contributes to life-time energy savings. Therefore, designers should seek first to maximise thermal performance in the building.	*PBL – Available as plasterboard laminate		

Health Comment	Rank	Environmental Issues	Rank	Cost Rank
Intact fully reacted polymers present no significant risk to occupants. Where ventilation is poor and volatile components are released, concentrations will build up and may affect occupants thereby elevating the rankings to 2/0.	0/0	High embodied energy. Pollution from manufacture and eventual disposal. Older materials in existing buildings is likely to contain CFCs.	2/0/0/2	784
Intact fully reacted polymers present no significant risk to occupants. Where ventilation is poor and volatile components are released, concentrations will build up and may affect occupants thereby elevating the rankings to 2/0.	0/0	High embodied energy. Pollution from manufacture and eventual disposal. Older materials in existing buildings is likely to contain CFCs.	2/0/0/2	876
Intact fully reacted polymers present no significant risk to occupants. Where ventilation is poor and volatile components are released, concentrations will build up and may affect occupants thereby elevating the rankings to 2/0.	0/0	High embodied energy. Pollution from manufacture and eventual disposal. Older materials in existing buildings is likely to contain CFCs.	2/0/0/2	136
Intact fully reacted polymers present no significant risk to occupants. Where ventilation is poor and volatile components are released, concentrations will build up and may affect occupants thereby elevating the rankings to 2/0.	0/0	High embodied energy. Pollution from manufacture and eventual disposal. Does not contain CFCs.	2/0/0/2	133
So long as fibres are contained, there is no risk to occupants foreseen. In the event of inhalable fibres being liberated into occupied areas, especially during maintenance, this risk may be elevated to 0/2.	0/0	Resin Binders. Pollution during manufacture and disposal.	1/0/0/2	100
So long as fibres are contained, there is no risk to occupants foreseen. In the event of inhalable fibres being liberated into occupied areas, especially during maintenance, this risk may be elevated to 0/2.	0/0	Resin Binders. Pollution during manufacture and disposal.	1/0/0/2	122

Application 2.5
TIMBER FRAMED WALL INSULATION AND CAVITY BARRIERS

Typical Situation

- plywood sheathing
- vapour barrier (polythene)
- insulation between timber framing
- plasterboard lining
- concrete floor

D.P.C.

Technical Requirements

Rot-resistant insulating material. Resistance to moisture absorption advantageous. Non-combustible material desirable and necessary in party walls. Provides essential part of sound insulation of party wall in timber systems.

Decay and Degradation Factors

Electrical and plumbing alterations can cause access to insulation both by building operatives and DIYs. Interstitial condensation risk with sub-standard workmanship. Fire. Chemical reaction with PVC cable/wiring. Frost and vermin.

Guidance Notes

Although the insulation receives protection from the plasterboard lining in the event of a fire it is still desirable to use a non-combustible insulation in this application and therefore technically mineral and glass wool seem most suitable. Good workmanship is essential with this type of construction to ensure effective installation of the vapour barrier and thus eliminate potential condensation problems. The vapour barrier if properly fixed should virtually eliminate fibre ingress to the interior. Quilted material is easier to fix than boards.

The health rating for mineral fibre is based upon the assumption that inhalable fibres are contained and for the plastic based products that there is sufficient ventilation to remove off-gassing contaminants in internal air.

The upstream score for the plastic based materials represents a higher embodied energy but, with housing, where buildings are normally long-lived it is important to note that the environmental impact of insulation will be relatively minor when considered against the reduction in impact that the material contributes to life-time energy savings. Therefore, designers should seek first to maximise thermal performance in the building.

From an environmental standpoint, man-made mineral fibre has a marginal edge over expanded polystyrene which in turn is marginally better than the other plastic products.
Cellulose fibre is manufactured from recycled paper and offers very little impact.

Alternatives	Technical Comment	Rank
A INSULATION 1 Mineral fibre (glass wool and rockwool)	Available as quilt or board. Has major advantage of non-combustibility in this application.	1
2 Expanded polystyrene	Combustible, but material is protected by plasterboard lining. Available with various fire properties.	3
3 Polyurethane foam	Combustible but material is protected by plasterboard lining. Available with various fire properties.	3
4 Polyisocy-anurate foam	Combustible - slightly better fire properties than other foamed plastics except phenolic and material is protected by plasterboard lining. Available with various fire properties.	3
5 Phenolic foam	Combustible, but burns with difficulty. Little smoke and toxic fume. Best fire properties of plastic foams.	2
6 Extruded polystyrene	Combustible, but material is protected by plasterboard lining. Available with various fire properties.	3
7 Cellulose fibre	Combustible, but material is protected by plasterboard lining. Available with various fire properties.	3
B CAVITY BARRIERS 1 Mineral fibre glass-wool or rockwool	Marginally easier to install than B2.	1
2 Softwood Timber	May require cutting to fit.	2

Health Comment	Rank	Environmental Issues	Rank	Cost Rank
So long as fibres are contained, there is no risk to occupants foreseen. In the event of inhalable fibres being liberated into occupied areas, especially during maintenance, the risk may be elevated to 0/2.	0/0	Resin Binders. Pollution during manufacture and disposal.	1/0/0/2	252*
Intact fully reacted polymers present no significant risk to occupants. Where ventilation is poor and volatile components are released, concentrations will build up and may effect occupants thereby elevating the rankings to 2/0.	0/0	High embodied energy. Pollution from manufacture and eventual disposal. Does not contain CFCs.	2/0/0/1	100*
Intact fully reacted polymers present no significant risk to occupants. Where ventilation is poor and volatile components are released, concentrations will build up and may effect occupants thereby elevating the rankings to 2/0.	0/0	High embodied energy. Pollution from manufacture and eventual disposal. Older material in existing buildings is likely to contain CFCs.	2/0/0/2	588*
Intact fully reacted polymers present no significant risk to occupants. Where ventilation is poor and volatile components are released, concentrations will build up and may effect occupants thereby elevating the rankings to 2/0.	0/0	High embodied energy. Pollution from manufacture and eventual disposal. Older material in existing buildings is likely to contain CFCs.	2/0/0/2	657*
Intact fully reacted polymers present no significant risk to occupants. Where ventilation is poor and volatile components are released, concentrations will build up and may effect occupants thereby elevating the rankings to 2/0.	0/0	High embodied energy. Pollution from manufacture and eventual disposal. Older material in existing buildings is like to contain CFCs.	2/0/0/2	123*
Intact fully reacted polymers present no significant risk to occupants. Where ventilation is poor and volatile components are released, concentrations will build up and may effect occupants thereby elevating the rankings to 2/0.	0/0	High embodied energy. Pollution from manufacture and eventual disposal. Older material in existing buildings is likely to contain CFCs.	2/0/0/2	161*
No health hazard foreseen for occupants so long as bulk material is contained. Too little is known of the material to be sanguine. If there is a possibility of exposure to substantial amounts of inhalable cellulose fibre, the risk may be elevated to 0/2.	0/0	Slight risk from Boron preservative. Low embodied energy.	0/0/0/0	N/A
No risk foreseen to occupants so long as the material remains contained as at manufacture.	0/0	Resin Binders. Pollution during manufacture. Very small quantities therefore minimal disposal.	1/0/0/1	119*
Virtually all woods used in construction are innocuous in situ. In this situation there is negligible risk from wood treatment and preservatives.	0/0		0/0/0/0	236*

* supply only

Application 2.6 PITCHED ROOF INSULATION	Alternatives	Technical Comment	Rank
Typical Situation insulation between structural members eaves ventilator ventilation ceiling 100mm cavity	1 Mineral wool quilt	Non-combustible. Material compacts over time thereby reducing insulation value. Mesh support between rafters.	1
	2 Loose mineral fibre (ceiling only)	Non-combustible. Sprayed application.	2
Technical Requirements Rot-proof insulation for laying over ceiling between joists or between rafters under felt and tiles. Water resistance and non-combustibility an advantage. Must be dimensionally stable.	3 Mineral fibre board/ slabs	Non-combustible. Cutting and fitting required.	3
Decay and Degradation Factors Wild life. Interstitial condensation. Fire. Possible contact with occupants' roof space access. Turbulence due to effective ventilation required to prevent condensation. Chemical reaction with PVC wiring/cables. **Guidance Notes** The technical choice between the alternatives depends upon the importance ascribed to non-combustibility of the insulation, but is not mandatory in the domestic situation. Good workmanship is essential with all alternatives to prevent cold bridging at eaves and at other connection details. It is essential to seal the ceiling and all holes/ducts entering the roof space to prevent the loss of fibres or loose fill and also to reduce the ingress to the habitable spaces otherwise there are health implications from those products manufactured using fibrous material. Gaskets and prefabricated hatches reduce the risk. Sealing also reduces risk from interstitial condensation.	4 Loose cellulose fibre (ceiling only)	Boron supplies, insecticidal and fungicidal protection for long-term durability. Combustible. Available with various fire properties.	3
No health risk to the occupiers is foreseen from the plastic products, but if combustability is an issue this has to be balanced with the health risk of other alternatives. The likelihood of contamination of the water supply via uncovered water storage tanks from dust and fibres should also be considered and appropriate action taken.	5 Vermiculite (ceiling only)	Can absorb moisture, lowering insulation property. Poor level of insulation for similar thickness of alternative material Sand dune effect from strong air movement in roof space. Non-combustible. (Rarely used.)	4
The upstream score for the plastic based materials represents a higher embodied energy but, with housing, where buildings are normally long-lived it is important to note that the environmental impact of insulation will be relatively minor when considered against the reduction in impact that the material contributes to life-time energy savings. Therefore, designers should seek first to maximise thermal performance in the building. From an environmental standpoint, man-made mineral fibres are marginally better than other alternatives with the exception of cellulose fibre which has little impact.	6 Polyisocy-aurate board	Combustible, but slightly better fire properties than other foamed plastics except phenolic and protected by class O lining and non-combustible covering. Adequate bending strength required to support roof fixers when boards fixed over rafters.	3

Health Comment	Rank	Environmental Issues	Rank	Cost Rank
Contained fibre presents no foreseeable hazard. In the loft situation there is a risk of contamination of occupied spaces and to the unprotected persons engaged in maintenance or disturbance activities. The B score is related to the amount of disturbance and risk of exposure and in extreme cases may be elevated to 3.	0/1	Resin binders. Pollution during manufacture and disposal.	1/0/0/2	178*
Contained fibre presents no foreseeable hazard. In the loft situation there is a risk of contamination of occupied spaces and to the unprotected persons engaged in maintenance or disturbance activities. The B score is related to the amount of disturbance and risk of exposure and in extreme cases may be elevated to 3.	0/1	Resin binders. Pollution during manufacture and disposal.	1/0/0/2	100*
Contained fibre presents no foreseeable hazard. In the loft situation there is a risk of contamination of occupied spaces and to the unprotected persons engaged in maintenance or disturbance activities. The B score is related to the amount of disturbance and risk of exposure and in extreme cases may be elevated to 3.	0/1	Resin binders. Pollution during manufacture and disposal.	1/0/0/2	176*
No hazard to occupants subject to no exposure to inhalable fibres. A definitive safety rating of this material awaits full toxicity testing. The addition of fire retardants, insecticides and fungicides will compound the problem. Blown fibre installation, lift ventilation, subsequent maintenance and loft access, and hiatuses in the ceiling expost tradesman, DIYers and occupants. The B score is related to the amount of disturbance and risk of export and in extreme may be elevated to 3.	0/1	Slight risk from Boron preservative. Low embodied energy.	0/0/0/0	N/A
In the absence of mineral fibre inclusions, no significant risk to occupants foreseen. If inhalable fibre is present and is released, the risk ranking may be elevated. Older materials may have asbestos or other fibrous content.	0/0	Natural material impacts from extraction, transport and processing.	1/0/0/0	176*
No significant risk to occupants foreseen.	0/0	High embodied energy. Pollution from manufacture and eventual disposal. Older materials in existing buildings is likely to contain CFCs.	2/0/0/2	1363*

* supply only

Application 2.6 (continued)	Alternatives	Technical Comment	Rank
	7 Extruded Polytstyrene Board	Combustible, but protected by class O lining and non-combustible covering. Adequate bending strength required to support roof fixers when boards fixed over rafters.	3
	8 Phenolic foam board	Combustible, but burns with difficulty. Little smoke and toxic fume. Can be made to satisfy class O therefore may be used without inner lining.	3
	9 Poly-urethane foam board	Combustible, may be protected by class O lining and non-combustible covering. Adequate bending strength required to support roof fixers when boards fixed over rafters.	3
	10 Expanded polystyrene board	Combustible, may be protected by class O lining and non-combustible covering. Adequate bending strength required to support roof fixers when boards fixed over rafters.	3
	11 Spray poly-urethane foam to underside rafters	Remedial treatment for securing tiles/slates suffering nail sickness.	N/A

Health Comment	Rank	Environmental Issues	Rank	Cost Rank
No significant risk to occupants foreseen.	0/0	High embodied energy. Pollution from manufacture and eventual disposal. Older materials in existing buildings is likely to contain CFCs.	2/0/0/2	156*
No significant risk to occupants foreseen.	0/0	High embodied energy. Pollution from manufacture and eventual disposal. Older materials in existing buildings is likely to contain CFCs.	2/0/0/2	229*
No significant risk to occupants foreseen.	0/0	High embodied energy. Pollution from manufacture and eventual disposal. Older materials in existing buildings is likely to contain CFCs.	2/0/0/2	1220*
No significant risk to occupants foreseen.	0/0	High embodied energy. Pollution from manufacture and eventual disposal. Does not contain CFCs.	2/0/0/2	208*
No significant risk to occupants foreseen.	0/0	High embodied energy. Pollution from manufacture and eventual disposal. Older material in existing buildings is likely to contain CFCs. May impair recycling of tiles or slates.	2/0/0/2	N/A

* supply only

Application 2.7
FLAT ROOF INSULATION

Typical Situation

WARM ROOF

solar protection
3 layer felt
insulation (rigid board)
vapour barrier

deck (concrete, steel, aluminium, woodwool)

solar protection
3 layer felt
timber deck
insulation
vapour check
ceiling

COLD ROOF

Technical Requirements
Rot proof insulation for laying in or on flat roof weathering systems. Water resistance an advantage – a necessity in inverted roof. Must be dimensionally stable in warm roof.

Decay and Degradation Factors
Wild life. Fire. Frost. Intestitial condensation. UV radiation. Maintenence actvities.

Guidance Notes
The risk from fire results from either burning debris from adjoining buildings on fire or from the room below. For the former, protection is provided from chippings used primarily for solar protection. For the latter, plasterboard lining normally provides the protection at ceiling level. Nevertheless non-combustible materials are preferred especially on metal decks exposed to the interior. Foams in this location could ignite due to heating from a fire below. Foams and cork are available 'cut to falls' facilitating provision of roof drainage. Extruded polystyrene foam board should be isolated from the membrane using woodfibre board in warm roof construction due to excessive shrinkage which causes stress in the membrane. Extruded polystyrene board has become a standard choice for inverted roof applications due to its good performance. Cellular glass is the only effective alternative with a closed cell, waterproof structure, but at a considerable cost premium. Mineral fibre board is available using a waterproof adhesive bonding making it suitable for this application.

The durability rating is related to the insulation material on its own, but if the waterproof membrane used has a shorter life expectancy, it is likely that the insulation will be replaced at the same time – especially in warm roof construction. With cold roofs very loose fibre fill it is essential to seal all holes/ducts entering the roof space to prevent ingress to the habitable spaces, otherwise there may be health implications from these materials. Sealing also reduces the risk from condensation.

The upstream score for the plastic based materials represents a higher embodied energy but, with housing, where buildings are normally long-lived it is important to note that the environmental impact of insulation will be relatively minor when considered against the reduction in impact that the material contributes to life-time energy savings. Therefore, designers should seek first to maximise thermal performance in the building.

From an environmental standpoint, corkboard, woodfibre and perlite offer least impact.

Alternatives	Technical Comment	Rank
1 Polyurethene foam board (warm roof)	Combustible. Large thermal movement – stabilize with glass tissue and partially bond membrane. Use HD grade.	2
2 Polyisocyanurate foam board (warm roof)	Combustible but slightly better fire properties than other foamed plastics except phenolic. Large thermal movement – stabilize with glass tissue and partially bond membrane. Use HD grade.	1
3 Expanded polystyrene beadboard (warm roof)	Combustible. Large thermal movement – separate membrane by layer of woodfibre or perlite. Use HD grade.	2
4 Phenolic foam	Combustible, but burns with difficulty. Best fire properties of foamed plastics.	1
5 Corkboard (warm roof)	Combustible. Low thermal movement. Needs additional support over troughed metal decking.	3
6 Woodfibre board (warm roof)	Combustible. Low thermal movement. Loses strength when wetted, subsequent decay unless preservative treated.	5
7 Mineral and glass fibre board (warm roof and inverted)	Non-combustible. Some types have low laminar strength, use mechanical fixing to deck.	1
8 Perlite (warm roof)	Non-combustible. Loses strength when wetted. Low laminar strength, use mechanical fixing as 7.	4
9 Cellular glass (inverted roof)	Non-combustible. Very low thermal movement.	1
10 Extruded polystyrene (warm roof and inverted)	Combustible. Lap joints minimise loss of insulation due to board shrinkage. Use super high density grade. Protection from UV radiation necessary (by ballast in inverted roof configurations).	2

Health Comment	Rank	Environmental Issues	Rank	Cost Rank
No significant risk foreseen to occupants.	0/0	High embodied energy. Pollution from manufacture and eventual disposal. Older materials in existing buildings is likely to contain CFCs.	2/0/0/2	162
No significant risk foreseen to occupants.	0/0	High embodied energy. Pollution from manufacture and eventual disposal. Older materials in existing buildings is likely to contain CFCs.	2/0/0/2	181
No significant risk foreseen to occupants.	0/0	High embodied energy. Pollution from manufacture and eventual disposal. Does not contain CFCs.	2/0/0/1	131
No significant risk foreseen to occupants.	0/0	High embodied energy. Pollution from manufacture and eventual disposal. Older materials in existing buildings is likely to contain CFCs.	2/0/0/1	144
No significant risk foreseen to occupants.	0/0	Low embodied energy. Durability assumes application in dry conditions. May be incinerated for disposal.	0/0/0/1	144
No significant risk foreseen to occupants.	0/0	Low embodied energy. Chemical additives may be used to resist moisture. May be incinerated for disposal.	0/0/1/1	100
No significant risk foreseen to occupants from intact product. Maintenance or DIY work may lead to increased risk from disturbance.	0/1	Resin binders. Pollution during manufacture and disposal.	1/0/0/2	149
No significant risk foreseen to occupants.	0/0	Inert natural material impacts from extraction transport and processing.	1/0/0/0	114
No significant risk foreseen to occupants.	0/0	High embodied energy. Pollution from manufacture. Potential for recycling depending upon method of fixing.	2/0/0/1	N/A
No significant risk foreseen to occupants.	0/0	High embodied energy. Pollution from manufacture and eventual disposal. Older materials in existing buildings is likely to contain CFCs.	2/0/0/2	131

Application 2.8
PIPE INSULATION

Typical Situation

insulation tape

formed sections

Technical Requirements

Hot and cold water supply: To provide sufficient insulation to prevent freezing of water in pipework in roof and other exposed locations.
Central heating pipework: Reduce heat loss from pipework throughout dwelling and prevent frost damage when system out of use.
Both types: Water resistant and rot-proof material essential. Good fire properties (i.e. resistance to ignition from small sources and resistance to flame spread) an advantage.

Decay and Degradation Factors

Wild life. Fire. Wear and tear. Mainly concealed in ducts, floor and roof spaces, but may be exposed to the interior.
Contamination of water supply via uncovered water storage tank by dust particles and fibres.

Guidance Notes

All types are satisfactory, but pre-formed sections are preferred due to the ease of application especially in confined spaces. Choice really depends upon cost factors plus additional consideration of appearance if the insulation is visible in interior spaces. In this circumstance, consideration might be given to the toxic fumes given off from the plastic types in the event of fire. This would only be a problem if a significant amount is fitted in an exposed position otherwise the problem is further reduced where the pipes are fixed in the usual position in the roof, floor and duct spaces.

From a health standpoint there, no significant risk to the occupants is foreseen, except in the case of the exposed plastic types in the event of fire or where fibres become liberated from the main-made mineral fibre types which may result in an elevation of risk.

Environmentally, mineral fibres material have a marginal advantage over the plastic varieties.

Asbestos fibre/magnesium lagging composition was commonly used for heating systems up to the late 1960s. In larger houses and blocks of flats the heating system may well have been insulated with asbestos lagging. **This material is potentially extremely dangerous and if suspected specialist advice should be sought on removal and disposal. See chapter 7**

Alternatives	Technical Comment	Rank
1 Mineral wool band or pipe wrap	Awkward to fix in confined spaces. Normally only used in concealed locations for frost protection. Good fire properties. Some varieties non-combustible.	2
2 Mineral wool formed sections	Formed sections usually thicker than mineral wool band therefore greater level of insulation. Good fire properties. Some varieties non-combustible.	1
3 Expanded polystyrene formed sections*	Combustible, available with various fire properties.	1
4 Isocyanurate formed sections*	Combustible, but slightly better fire properties than other foamed plastic except phenolic. Available with various fire properties.	1
5 Foamed nitrile/ synthetic formed sections*	Usually complete tube permitting convenient fixing on new work. Combustible, available with various properties.	1
6 Extruded polystyrene formed sections*	Combustible, available with various fire properties.	1
7 Polyurethane formed sections*	Combustible – available with various fire properties.	1
8 Phenolic foam formed sections*	Combustible, but burns with difficulty. Little smoke and toxic fumes. Best fire properties of plastic foams.	1

* products of combustion are toxic

Health Comment	Rank	Environmental Issues	Rank	Cost Rank
No significant risk is foreseen to occupant while the product is intact. With maintenance and DIY, particularly when the resin binder is embrittled by age and heat, inhalable fibres will become liberated.	0/2	Resin bonded. Pollution during manufacture and disposal.	1/0/0/2	251
No significant risk is foreseen to occupant while the product is intact. With maintenance and DIY, particularly when the resin binder is embrittled by age and heat, inhalable fibres will become liberated.	0/2	Resin bonded. Pollution during manufacture and disposal.	1/0/0/2	145*
No significant risk is foreseen to occupant.	0/0	High embodied energy. Pollution from manufacture and eventual disposal. Does not contain CFCs.	2/0/0/2	100*
No significant risk is foreseen to occupant.	0/0	High embodied energy. Pollution from manufacture and eventual disposal. Older material in existing buildings is likely to contain CFCs.	2/0/0/2	N/A
No significant risk is foreseen to occupant.	0/0	Pollution from manufacture. Downstream score 1 if disposed of by combustion.	2/0/0/0	N/A
No significant risk is foreseen to occupant.	0/0	High embodied energy. Pollution from manufacture and eventual disposal. Older material in existing buildings is likely to contain CFCs.	2/0/0/2	N/A
No significant risk is foreseen to occupant.	0/0	High embodied energy. Pollution from manufacture and eventual disposal. Older material in existing buildings is likely to contain CFCs.	2/0/0/2	204*
No significant risk is foreseen to occupant.	0/0	High embodied energy. Pollution from manufacture and eventual disposal. Older material in existing buildings is likely to contain CFCs.	2/0/0/2	N/A

* supply only

Application 2.9 HOT & COLD WATER TANK INSULATION	Alternatives	Technical Comment	Rank
Typical Situation cylinder jacket encased in polythene bag rigid board insulation box filled with loose insulation	1 Mineral wool jackets	Easy to fix. Non-combustible insulation encased in combustible jacket. Intended for hot water cylinders.	1
	2 Mineral wool quilt	Non-combustible. Assessment assumes roof space application.	1
	3 Mineral fibre board	Easy to cut and fix around rectangular tank. Non-combustible. Assessment assumes roof space application.	1
	4 Expanded polystyrene board	Easy to cut and fix around rectangular tank. Combustible, available with various fire properties.	1
	5 Polyurethane board	Easy to cut and fix around rectangular tank. Combustible, available with various fire properties.	1
	6 Isocyanurate board	Easy to cut and fix around rectangular tank. Combustible, but slightly better than other foamed plastics. Available with various fire properties.	1
	7 Factory preformed polystyrene/ polyurethane foam pre-attached to tank (hot water only)	Very convenient for new applications. Combustible, available with various fire properties.	1

Technical Requirements

Hot water cylinders – to reduce heat loss. Cold water storage tanks – to prevent freezing. Both applications: water resistant, rot-proof material essential. Dimensional stability an advantage. Good fire properties (i.e. resistance to ignition, non-combustibility) an advantage.

Decay and Degradation Factors

Vermin and wild life in roof space. Fire. Wear and tear. Easily accessible in airing cupboards where long-term contact with occupants' clothing is likely. Contamination of water supply via uncovered water storage tanks by dust particles and fibres.

Guidance Notes

When the hot cylinder is fitted in the usual position in an airing cupboard the standard mineral wool jacket or pre-foamed polyurethane lagged cylinder have the main advantage of ease of fitting. For rectangular cold water tanks board types are generally easier to fit, as are jackets for circular tanks. There may be problem locations where loose fill of other material (eg. Plywood) inside a container might prove most suitable.

It is essential that open tanks have a close fitting (not sealed) lid to reduce contamination of stored water as this water may be used for drinking. Board type insulation should not be relied upon as the only covering for open tanks. Combustible materials are susceptible to ignition and care is required during plumbing operations when using a naked flame. Board systems need fixing tapes.

No significant risk to the occupants is foreseen from any of the alternatives except in the case of fibre release from the M.M.M.F. fibre boards or jackets where an elevation of risk may occur possibly to unacceptable levels.

From an environmental standpoint, the M.M.M.F. fibre boards have a marginal advantage over the plastic types.

Health Comment	Rank	Environmental Issues	Rank	Cost Rank
Contained fibre presents no foreseeable hazard. In the event of inhalable fibres being liberated into occupied areas, especially during maintenance, the risk may be elevated to 0/2.	0/0	Resin bonded. Pollution during manufacture and disposal.	1/0/0/2	100*
Contained fibre presents no foreseeable hazard. In the event of inhalable fibres being liberated into occupied areas, especially during maintenance, the risk may be elevated to 0/2.	0/0	Resin bonded. Pollution during manufacture and disposal.	1/0/0/2	102*
Contained fibre presents no foreseeable hazard. In the event of inhalable fibres being liberated into occupied areas, especially during maintenance, the risk may be elevated to 0/2.	0/0	Resin bonded. Pollution during manufacture and disposal.	1/0/0/2	578*
No significant risk foreseen to occupants.	0/0	High embodied energy. Pollution from manufacture and eventual dispersal. Does not contain CFCs.	2/0/0/1	430*
No significant risk foreseen to occupants.	0/0	High embodied energy. Pollution from manufacture and eventual disposal. Older materials in existing buildings is likely to contain CFCs.	2/0/0/2	630*
No significant risk foreseen to occupants.	0/0	High embodied energy. Pollution from manufacture and eventual disposal. Older materials in existing buildings is likely to contain CFCs.	2/0/0/2	269*
No significant risk foreseen to occupants.	0/0	High embodied energy. Pollution from manufacture and eventual disposal. Older materials in existing buildings is likely to contain CFCs. As 4 above if polystyrene.	2/0/0/2	Included in the cost of the tank

* supply only

Application 3.1 WINDOW/EXTERNAL DOOR FRAMES	Alternatives	Technical Comment	Rank
Typical Situation DOOR FRAME–PLAN SECTION 	1 Timber (a) Softwood	Can be durable solution if frame vacuum impregnated with preservative prior to installation.	2
	(a) temperate hardwood	Can be durable without other treatments.	1
	(c) tropical hardwood	Can be durable without other treatments.	1
Technical Requirements Framing system of adequate strength and good appearance to support glazing and/or doors as a means of fixing within the structural opening of a wall. Self decoration or ability to receive decoration an advantage. System must include means of excluding passage of water and air.	2 Unplasticised Polyvinyl Chloride (PVCu)	For larger frames needs reinforcement using galvanised steel sections. Self finished.	1
Decay and Degradation Factors Normal weathering. Frost. Ultra-violet light. Normal wear and tear. Vandalism and intruders. Wildlife. Cleaning and cleaning solutions. Abrasion for re/decoration.	3 Aluminium	Self finished using anodised coatings. Cold bridge effect unless superior (and more expensive) 'thermal break' section utilised. When ranking would be 1.	2
Guidance Notes From an environmental perspective softwood and temperate hardwood offer the best compromise. In the past tropical hardwood has been selected on grounds of appearance and durability, but for reasons explained in Chapter 3 use of tropical hardwoods should be discontinued. Use of gasket systems or other forms of air sealing is recommended to improve air tightness and therefore thermal performance. Gasket sealing is usually standard in alternatives 2-4 and achieved by high-performance standard timber frames. Timber systems involve regular maintenance cycles.	4 Galvanised Steel	Slender cross section possible due to high strength of steel. Powder coating provides self finish. Cold bridge effect.	2

Health Comment	Rank	Environmental Issues	Rank	Cost Rank
No significant risk foreseen to occupants.	0/0	Available from sustainable sources. Life expectancy related to maintenance and preservative treatments – see applications Paint and Wood Preservatives, 7.7, 7.9 and 7.10. Disposal by combustion causes atmospheric pollution.	1/0/1/0	131
No significant risk foreseen to occupants.	0/0	Availability from sustainable sources is limited. Upstream score 1 if from sustainable managed sources.	2/0/0/0	357
No significant risk foreseen to occupants.	0/0	Large proportion of tropical timber is still sourced from virgin forest. Upstream score 1 if from sustainable managed sources.	3/0/0/0	321
No significant risk foreseen to occupants.	0/0	Problems from pollution during manufacture – but situation is improving. Problems at disposal. Galvanised reinforcement sections in PVC frames is similar to 4 (below).	2/0/1/2	384
No significant risk foreseen to occupants.	0/0	High embodied energy. Pollution from manufacture. High recycled content. Energy loss through cold bridging if not thermally broken. Coatings may cause problems with recycling.	2/0/1/1	241
No significant risk foreseen to occupants.	0/0	Steel could be recycled, but finishes may cause problems. Impacts from powder coating not known. Energy loss through cold bridging if not thermally broken.	2/0/1/1	100

Application 3.2 LINTELS AND ARCHES	Alternatives	Technical Comment	Rank
Typical Situation GALVANISED STEEL REINFORCED CONCRETE/ ARTIFICIAL STONE VERTICAL SECTION	1 Reinforced concrete and artificial stone	Boot lintel – serious cold bridge. Two separate lintels are better in cavity walls. Ranking based on this configuration.	2
	2 Galvanised steel	Lightweight propriety prefabricated units supports both leaves of cavity wall with integral d.p.c.	1
	3 Steel	Hot rolled sections. Needs corrosion protection.	4
Technical Requirements Adequate durable means of support over door and window openings. Must dimensionally co-ordinate with brick and block sizes. Integral means of closing the cavity and damp-proofing over opening is seen by some as an advantage. Should not reduce thermal insulation properties of the wall or introduce cold bridge.	4 Softwood	Can be a durable solution if vacuum impregnated with preservative prior to installation in external wall. Reduces cold briding.	2
Decay and Degradation Factors Normal weathering. Frost. Differential movement. Run off from masonry cement and lime mortars. Fungal and insect attack (timber). Fire.	5 Temperate hardwood	Can be durable without other treatments. Reduces cold bridging.	1
	6 Tropical hardwood	Can be durable without other treatments. Reduces cold bridging.	1
Guidance Notes Selection of lintels and arches is influenced by aesthetic criteria, the main choice being whether or not the appearance of a visible structural lintel is desirable. Options 2, 3, 8 and 9 can provide a 'no-lintel' appearance. Arch effects are now commonly achieved using curved lintels, such as options 2 or 8.	7 Natural stone	Limited span for normal cross section. Heavy weight. Durability related to source.	2
Technical selection is based upon required span, thermal performance, durability and buildability. Natural stone and timber will require larger cross-section or additional support over larger openings. Timber reduces cold-bridging effects, as can appropriate detailing in insulated cavity wall construction. From a durability perspective there is not a great deal to choose between options provided adequate detailing and manufacturers advice is followed, eg. direction of bedding layers in sandstone, preservative treatments of softwoods, etc. Buildability issues, such as weight, manhandling, ease of installation (eg. Options 2, 4-8), as well as integral cavity tray systems (options 2 and 8) are all an advantage.	8 Stainless steel	Lightweight propriety prefabricated units supports both leaves of cavity wall with integral d.p.c.	1
Environmentally recycled lintel components represent a good option, otherwise 1, 3, 4, 7 and 9 have a slightly lower environmental impact. Use of tropical hardwood should be avoided – see Chapter 3.	9 Reinforced bricks/block	Stainless steel mesh laid in brick or blockwork joints or propriety lintel blocks as permanent formwork for in situ reinforced concrete. Both systems require propping.	3

Health Comment	Rank	Environmental Issues	Rank	Cost Rank
No significant risk foreseen to occupants.	0/0	Moderate embodied energy. Reinforcement steel inhibits recycling.	1/0/0/1	168*
No significant risk foreseen to occupants.	0/0	High embodied energy. Cold bridging. Finishes inhibit recycling.	2/0/0/1	299
No significant risk foreseen to occupants.	0/0	High embodied energy. Durability related to protection from dampness.	2/0/1/0	244
No significant risk foreseen to occupants.	0/0	Available from sustainable sources. Life expectancy related to maintenance and preservative treatments – see applications Paints and Wood Preservatives, 7.7, 7.9 and 7.10. Can be recycled.	1/0/1/0	100
No significant risk foreseen to occupants.	0/0	Availability from sustainable sources is limited. Upstream score 1 if from sustainable managed sources.	2/0/0/0	395+
No significant risk foreseen to occupants.	0/0	Large proportion of tropical timber is still sourced from virgin forest. Upstream score 1 if from sustainable managed sources.	3/0/0/0	215++
No significant risk foreseen to occupants.	0/0	Quarrying. Recycling potential.	1/0/0/0	1020
No significant risk foreseen to occupants.	0/0	Assumes UK source with 100% recycled content. Material from other sources increase upstream score to 2.	1/0/0/0	341
No significant risk foreseen to occupants.	0/0	Moderate embodied energy. Reinforcement steel inhibits recycling.	1/0/0/1	N/A

* Concrete
+ American Oak
++ Iroko

Application 3.3
CILLS AND THRESHOLDS

Typical Situation

VERTICAL SECTION

Technical Requirements

Durable means of closing cavity wall under door or window openings that includes method of shedding water from the wall. Self finish or self decoration an advantage. Should not reduce thermal properties of the wall or introduce cold bridge.

Decay and Degradation Factors

Normal weathering. Frost. Run off from masonry cement and lime mortars. Impact. Vandalism and intruders. Ultra-violet light. Acid rain. Cleaning and cleaning solutions. Foot traffic (thresholds), abrasion for decoration.

Guidance Notes

Selection is influenced by aesthetic criteria, particularly for window cills.

These applications represent the more extreme exposure conditions in the external wall, therefore selection for durability and water exclusion form the primary technical requirements. From this point of view the imperious materials offer an advantage (1, 2, 7, 9, 11 and 12). Timber based systems require regular maintenance.

From a buildability perspective heavier alternatives such as 1 and 6 may require lifting equipment for larger openings. Systems made up from 8 and 10 require setting in mortar with careful detailing and workmanship.

Environmentally recycled components represent a good option, otherwise 1, 3 and 6 have lower environmental impact. Use of tropical hardwood should be discontinued – see Chapter 3.

Alternatives	Technical Comment	Rank
1 Reinforced concrete and artificial stone	Heavy and difficult to install in larger openings.	1
2 Galvanised steel	Lightweight proprietary prefabricated sections. Powder coated self finish.	1
3 Softwood (cills only)	Can be a durable solution if vacuum impregnated with preservative prior to installation and with adequate maintenance.	4
4 Temperate hardwood	Can be durable without other treatments.	2
5 Tropical hardwood	Can be durable without other treatments.	2
6 Natural stone	Durability of stone, particularly in terms of frost protection. Needs careful consideration of detailing and stone type.	1-3
7 uPVC	Self finished lightweight propietary prefabricated sections.	2
8 Bricks/ blocks	Purpose-made units laid in situ. Mortar strength is important for overall durability. Ease of installation an issue.	3
9 Aluminium	Anodised self finished lightweight propietary prefabricated sections.	1
10 Clay/ concrete tiles	Purpose-made units laid in situ. Mortar strength is important for overall durability. Ease of installation an issue.	3
11 Lead	Purpose made, usually fixed on timber substrata – see softwood.	2
12 Zinc	Purpose made.	2

Health Comment	Rank	Environmental Issues	Rank	Cost Rank
No significant risk foreseen to occupants.	0/0	Moderate embodied energy. Reinforcement mesh inhibits recycling.	1/0/0/1	427+
No significant risk foreseen to occupants.	0/0	High embodied energy. Finishes inhibit recycling.	2/0/0/1	195++
No significant risk foreseen to occupants.	0/0	Available from sustainable sources. Life expectancy related to maintenance and preservative treatments – see application Paints and Wood Preservatives, 7.7, 7.9 and 7.10. Can be recycled.	1/0/1/0	284
No significant risk foreseen to occupants.	0/0	Availability from sustainable sources is limited. Upstream score 1 if from sustainably managed sources.	2/0/0/0	369
No significant risk foreseen to occupants.	0/0	Large proportion of tropical timber is still sourced from virgin forest. Upstream score 1 if from sustainably managed sources.	3/0/0/0	331
No significant risk foreseen to occupants.	0/0	Quarrying. Recycling potential.	1/0/0/0	1065
No significant risk foreseen to occupants.	0/0	Problems from pollution during manufacture – but situation is improving. Problems at disposal.	2/0/1/1	475
No significant risk foreseen to occupants.	0/0	Various types – also see Application 7.3.		201
No significant risk foreseen to occupants.	0/0	High embodied energy. Pollution from manufacture. High recycled content.	2/0/1/1	318
No significant risk foreseen to occupants.	0/0	Moderate to high embodied energy dependent on type.	2/0/1/1	100
No significant risk foreseen to occupants.	0/0	High-embodied energy. Pollution from manufacture. High recycled content.	2/0/0/0	311+++
No significant risk foreseen to occupants.	0/0	As lead but question percentage of recycled content.	2/0/0/0	N/A

+ cast stone
++ excludes powder coating
+++ on softwood cill

Application 3.4 GLAZING FIXING SYSTEMS	Alternatives	Technical Comment	Rank
Typical Situation	1 Linseed oil putties	Satisfactory for general domestic applications to painted frames (not suitable for double glazing).	3
	2 Synthetic rubber base glazing compound	Satisfactory for general domestic applications, but can also be used for unpainted timber frames protected by decorative wood stains.	1
	3 Polysulphide base glazing compound	Satisfactory for general domestic applications, but can also be used for unpainted timber frames protected by decorative wood stains.	1
	4 Self-adhesive glazing strip	Satisfactory and convenient systems for use with glazing beads.	1
	5 Beading		
	(a) timber	(a) Normally hardwood, essential for durability.	2
	(b) plastic PVCu	(b) Proprietary sections include gasket systems. Beading as part of window system.	1
	(c) aluminium	(c) Proprietary sections include gasket systems. Beading as part of window system. Simple sections can be used as beads in timber frames.	1

Technical Requirements

Effective means of conveniently fixing and sealing glazing securely with frames. Putties, etc, must harden sufficiently to hold glass but still remain sufficiently soft to remain an effective air and water seal and accommodate movements between glass and timber frame. Beading systems must be secured by glazing pins, screws or clips. Ability to receive decoration may be necessary.

Decay and Degradation Factors

Natural weathering. Frost action. Wild life. Abrasion (sanding) prior to repainting. Periodic replacement through damage to glass or normal ageing. Cleaning and cleaning solutions.

Guidance Notes

The glazing compounds and beading systems may be used in various combinations in the range of frame and glazing options. The main comments refer to double glazing which is assumed to be a minimum standard from energy and comfort requirements. Manufacturers recommend drained systems.

Technically and environmentally there is little to choose between modern compound options and strips (2-3). Strips are marginally easier to use with timber beads. Aluminium and PVCu windows use proprietry beads and gasket systems of good performance. The tropical hardwood used for beads is usually Ramin which is an endangered species and so use should be discontinued.

For single glazing into painted timber frames unleaded linseed oil putty will prove most appropriate and cost effective.

In older buildings it seems wise when undertaking maintenance work to consider that existing putty may have a lead content and to treat the material in a similar manner to lead-based paints as described in Chapter 8, i.e. avoiding dry sanding as a means of preparation or removal.

Health Comment	Rank	Environmental Issues	Rank	Cost Rank	
No significant risk foreseen to occupants. (Old leaded materials presented an ingestion hazard to infants, particularly in association with lead based paints.)	0/0	Small volume. Moderate energy. Assuming zero lead content.	0/0/1/1	Refer to specialist supplier	
No significant risk foreseen to occupants.	0/0	Small volume of material but high energy in extraction, refining and conversion.	1/1/1/1		
No significant risk foreseen to occupants.	0/0	Small volume of material but high energy in extraction, refining and conversion.	1/1/1/1		
No significant risk foreseen to occupants.	0/0	Small volume of material but high energy in extraction, refining and conversion, but likely to involve lower solvent release at point of application.	1/0/1/1		
No significant risk foreseen to occupants.	0/0	For detailed comments see Application 7.6, Timber. (a) Tropical hardwood Temperate hardwood (b) Problems from pollution during manufacture but situation is improving. Reusable at reglazing or resealing. (c) High embodied energy. Pollution from manufacture. Reusable at reglazing or resealing.	3/0/0/0 2/0/0/0 2/0/0/1 2/0/0/0		

Application 3.5 SEALANTS TO DOOR AND WINDOW FRAMES	Alternatives	Technical Comment	Rank
Typical Situation D.P.C. cavity wall 100mm cavity SEALANT plan glass window frame	Sealants usually based on one or more of the following:	Various formulation combinations are available from manufacturers. Generally satisfactory for domestic use. Life expectancy:	
	1 Acrylic	15 years	2
	2 Polysulphide	20 years	1
	3 Polyurethane	20 years	1
Technical Requirements Provide impervious draught seal between frames and structure Must be rot and frost resistant and cope with movement. Good appearance necessary or be able to take decoration. Capable of adhering to a wide range of materials, and must not slump at high temperatures.	4 Silicone	up to 30 years	1
	5 Oleo-resinous (oil)	5 to 10 years	3
Decay and Degradation Factors Normal weathering. Frost. Maintenance and decoration. Mould growth, vegetation and wild life. Cleaning and cleaning solutions. Acid rain.		Grading based on life expectancy, but other factors such as ease of application may influence choice.	

Guidance Notes

Most systems perform adequately during their normal life-expectancy. Durability is normally related to cost, i.e. the lower the cost the lower the life expectancy. Environmentally there is little to choose between systems except that it is clearly preferable to use more durable systems as they will require less material over the buildings' life.

In existing buildings, older formulations using asbestos as a filler or lead as a drying agent could pose a slight hazard but if one considers the small quantities used in the domestic situation the risk to the occupant is of a low order, provided dry sanding is not used as a means of preparation for decoration, or as a means of removal.

Health Comment	Rank	Environmental Issues	Rank	Cost Rank
				Refer to specialist supplier
No significant risk foreseen to occupants.	0/0	Small volume of material, but high energy in extraction, refining and conversion. Moderate life expectancy.	1/1/1/1	
No significant risk foreseen to occupants.	0/0	Small volume of material, but high energy in extraction, refining and conversion.	1/1/0/1	
No significant risk foreseen to occupants.	0/0	Small volume of material, but high energy in extraction, refining and conversion.	1/1/0/1	
No significant risk foreseen to occupants.	0/0	Small volume of material, but high energy in extraction, refining and conversion.	1/1/0/1	
No significant risk foreseen to occupants.	0/0	Small volume of material, but high energy in extraction, refining and conversion. Lower life expectancy.	1/1/2/1	

Application 3.6 FIRE DOORS	Alternatives	Technical Comment	Rank
Typical Situation 'HALF HOUR' DOOR Intumescent strip - needed some types of ½ hour door essential for longer periods of fire resistance.			
	1 Flaxboard core	Choice really depends upon finish and appearance required provided doors comply with BS 476 Part 22 clause 6, 7 & 8 for required period for fire resistance.	1
	2 Mineral wool fibre board core	Choice really depends upon finish and appearance required provided doors comply with BS 476 Part 22 clause 6, 7 & 8 for required period for fire resistance.	1
Technical Requirements Door and frame combination to provide ½ to 2 hour fire resistance (½ hour most common requirement with domestic applications). Construction of door together with appropriate ironmongery must provide adequate strength, racking and warp resistance for both normal usage and in the event of a fire. Combustible material permissible provided door satisfies BS fire test requirements. Quality of factory applied finish important or must accept decoration, using paints and/or varnishes.	3 Calcium silicate board core	Choice really depends upon finish and appearance required provided doors comply with BS 476 Part 22 clause 6, 7 & 8 for required period for fire resistance.	1
Decay and Degradation Factors Fire. Normal wear and tear, particularly impact damage. Cleaning and cleaning solutions. Abrasion (sanding down) prior to repainting.	4 Blockboard core	Choice really depends upon finish and appearance required provided doors comply with BS 476 Part 22 clause 6, 7 & 8 for required period for fire resistance.	1
Guidance Notes For the currently available alternatives technical selection depends upon appearance provided doors satisfy fire rating standards mentioned under technical comments. Environmentally natural fibre core (flaxboard) or reused door is to be preferred. Fire doors including asbestos insulating board were used extensively in the past. It is often difficult to identify doors with asbestos cores and although the material is contained by facings (usually plywood) and timber lippings. It is better to replace the door under the precautionary principle. Although it was unusual to use asbestos board as the finished surface on the majority of doors fixed in new construction, it was fairly common to upgrade existing panel doors in refurbishment schemes by adding asbestos board to one face. Doors subject to damage or foreseeable damage should be replaced by one of the other Alternatives. The existing asbestos material should be disposed of properly – see Chapter 7.	5 Asbestos insulation board	No longer available. Careful disposal of old doors – see guidance notes.	-

Health Comment	Rank	Environmental Issues	Rank	Cost Rank
		Environmental comment based on core material only, see general comments on applications Internal and External Doors, 3.7 and 3.8.	0/0/0/1	
No significant risk foreseen to occupants.	0/0	Natural fibre. Downstream core 0 if reused.	1/0/0/1	100+*
No significant risk foreseen to occupants so long as the door remains intact. In the event of damage that releases substantial numbers of inhalable fibres, the risks may be elevated to 0/2.	0/0	Resin binders. Pollution during manufacture. Downstream score 0 if reused.	2/0/0/1	140++*
No significant risk foreseen to occupants so long as the door remains intact. If the `calcium silicate' includes mineral fibres and damage releases substantial numbers of inhalable fibres, the risks may be elevated to 0/2.	0/0	Moderate embodied energy. Silicate from quarrying. Downstream score 0 if reused.	1/0/0/1	210++*
No significant risk foreseen to occupants.	0/0	Upstream score assumes softwood from sustainable sources. Resin adhesives. Downstream score 0 if reused.	-/-/-/3	140++*
Fibre release on ageing, cleaning, maintenance and disposal present risk to maintenance workers, DIY occupants, construction and demolition operatives. Doors are especially vulnerable to damage.	3/3	Asbestos is a hazardous waste and needs to be disposed of properly. Asbestos in existing buildings needs to be identified, recorded, assessed for comparative risks and removed or perhaps treated or encased and managed as appropriate. The (-) classification refers to lack of availability/no longer used.		N/A

* supply only
+ half hour
++ one hour

Application 3.7 EXTERNAL DOORS	Alternatives	Technical Comment	Rank
Typical Situation ELEVATION FRAME SECTION	1 Timber		
	(a) softwood	(a) Can be a durable solution if vacuum impregnated with preservative prior to fixing.	2
	(b) temperate hardwood	(b) Can be durable without other treatments.	1
	(c) tropical hardwood	(c) Ditto.	1
Technical Requirements High strength to resist intruders and vandalism. Ability to receive hinges and ironmongery, particularly security equipment. Sound insulation, thermal insulation. Resistance to moisture. Ability to take decoration. Appearance. Ease of cleaning.	2 Powder or plastic coated galvanised steel	Proprietary composite panel system. Can include sophisticated security systems.	1
Decay and Degradation Factors Normal weathering. Frost. Redecoration. Cleaning and cleaning solutions. Fungal attack. Normal wear and tear. Vandalism and intruders.	3 Unplasticised Polyvinyl Chloride (PVCu)	Either proprietary composite panel system. Can include sophisticated security systems or framed and glazed using gasket systems.	1
Guidance Notes A primary selection factor is appearance. Technically security considerations may predominate, but should be considered in combination with the locks and ironmongery. External doors to some dwellings, eg. Flats, may also need to be fire-doors, see Application 3.6. For normal applications thermal performance and air tightness are key parameters and can vary significantly between types and manufacturers. A good thermal standard is an environmental requirement and specifiers should select high performance door sets (ie. pre-fabricated door and frame systems) available in aluminium, PVCu, steel or timber. Overall softwood appears to have the lowest environmental impact including the thermal and airtightness criteria. Recycled doors also provide a low impact option. From a health perspective, glazed panels pose a safety risk. Laminated or tempered safety glass should be used. Use of tropical hardwood should be discontinued – see Chapter 3.	4 Medium Density Fibreboard	Proprietary composite panel system. Relatively new material.	2
	5 Aluminium	Usually framed and glazed using gasket systems.	1

Health Comment	Rank	Environmental Issues	Rank	Cost Rank
Exposure to substantial amounts of wood dust is associated with adverse effects on breathing, and with cancer but would only be of concern in DIY sanding. (a) No significant risk foreseen to occupants.	0/0	(a) Available from sustainable sources. Life expectancy related to maintenance and preservative treatments – see application Paint and Wood Preservatives, 7.7, 7.9 and 7.10. Disposal by combustion causes atmospheric pollution.	1/0/1/0	100
(b) No significant risk foreseen to occupants.	0/0	(b) Availability from sustainable sources is limited. Upstream score 1 if from sustainably managed sources.	2/0/0/0	138
(c) No significant risk foreseen to occupants.	0/0	(c) Large proportion of tropical timber is still sourced from virgin forest. Upstream score 1 if from sustainably managed sources.	3/0/0/0	125
No significant risk foreseen to occupants.	0/0	Steel could be recycled but finishes may cause problems. Impacts from powder coating not known.	2/0/1/1	208
No significant risk foreseen to occupants.	0/0	Problems from pollution during manufacture but situation is improving. Problems at disposal. Galvanised reinforcement similar to 2 above.	2/0/1/2	250*
Exposure to substantial amounts of wood dust is associated with adverse effects on breathing, and with cancer but would only be of concern in DIY sanding.	0/0	Upstream score 1 if fibre sourced from forest trimmings and waste. Life expectancy related to maintenance. Resin binder.	2/0/1/1	N/A
No significant risk foreseen to occupants.	0/0	High embodied energy. Pollution from manufacture. High recycled content. Coatings may cause problems with recycling.	2/0/1/1	526*

* door and frame

Application 3.8 INTERNAL DOORS	Alternatives	Technical Comment	Rank
Typical Situation TIMBER FLUSH DOOR S.W. lipping S.W. frame core FACING (plywood/hardboard or leaboard)	1 Timber framed (a) Softwood (b) Temperate hardwood (c) Tropical hardwood	 (a) Standard solution for framed and glazed or panel doors. Requires decoration. (b) Used only for decorative purposes. (c) Used only for decorative purposes.	 1 1 1
Technical Requirements Strength sufficient for internal purposes. Resist minor impacts. Ability to receive hinges and ironmongery. Sound insulation. Fire resistance may be required. Ease of decoration. Ease of cleaning. Appearance. Control of movement.	2 Composite: (a) Hardboard (various cores) (b) Blockboard (c) Plywood (various cores)	 (a) Can be moulded for decorative effects or panelling. Ranking dependant upon cost. (b) Very robust solid timber core. (c) Can be veneered for decorative purposes.	 2/3 1 1
Decay and Degradation Factors Normal usage (and abusage). Redecoration, cleaning and cleaning solutions. Fire.	3 Aluminium	Propietary systems for framed and glazed and sliding doors.	1
Guidance Notes The primary selection factor is appearance. Strength and sound performance and overall quality is usually related to cost. Softwood offers the lowest environmental impact and is available for most applications and in a variety of appearance/forms. Recycled doors also provide a low-impact option. Use of tropical hardwood should be discontinued – see Chapter 3. In existing buildings older paint films may contain lead which should be removed carefully, see Chapter 7. Glazed panels pose a safety risk. Laminated or tempered safety glass should be used. Disposal of timber based doors by combustion causes atmospheric pollution. Use of tropical hardwood should be discontinued. See chapter 3	4 Unplasticised Polyvinyl Chloride (PVCu)	Propietary systems for framed and glazed and sliding doors.	1
	5 Medium Density Fibreboard (MDF)	Can be moulded for decorative effects/special shapes.	1

Health Comment	Rank	Environmental Issues	Rank	Cost Rank
(a) – (c) No significant risk foreseen to occupants in the intact state. (Exposure to substantial amounts of fine wood dust is associated with adverse effects on breathing, and with cancer, but would only be of concern in DIY sanding.)	0/0 0/0 0/0	Disposal by combustion causes atmospheric pollution. (a) Availability from sustainable sources. (b) Availability from sustainable sources is limited. Upstream score 1 from sustainably managed sources (c) Large proportion of tropical timber is still sourced from virgin forest. Upstream score 1 if from sustainably managed sources.	1/0/0/0 2/0/0/0 3/0/0/0	114* 227* 204*
(a) – (c) No significant risk foreseen to occupants. (With poor ventilation the build up of acute irritant and carcinogenic vapours from resin binders may lead to elevated risk of 0/3.) See also note on sanding in 1(a)-(c)	0/0 0/0 0/0	Disposal by combustion causes atmospheric pollution. (a) Upstream score assumes softwood fibre from sustainable sources. (b) Plywood facings usually contain tropical hardwood and resin adhesive. (c) Plywood facings usually contain tropical hardwood and resin adhesive.	1/0/0/0 2/0/0/1 2/0/0/1	132* 309* 274*
No significant risk foreseen to occupants.	0/0	High embodied energy. Pollution from manufacture. High recycled content coatings may cause problems with recycling.	2/0/1/1	688*
No significant risk foreseen to occupants.	0/0	Problems from pollution during manufacture but situation is improving. Problems at disposal. Galvanised reinforcement.	2/0/1/2	867+*
No significant risk foreseen to occupants in the intact state. (Exposure to substantial amounts of fine wood dust is associated with adverse effects on breathing, and with cancer, but would only be of concern in DIY sanding.)	0/0	Upstream score 1 if fibre sourced from forest trimmings and waste. Life expectancy related to maintenance. Resin binder.	2/0/1/1	100*

* supply only
+ part glazed

Application 3.9 ## LEADED LIGHTS	Alternatives	Technical Comment	Rank
Typical Situation TRADITIONAL RATIONALIZED glass inside face lead framework soldered at joints lead came effect achieved by lead section glued to glass	1 Traditional lead	Generally used for stained glass and for replacement work on historic and other buildings of character.	1
	2 Traditional copper	Could be used as an alternative to traditional lead, but not generally available.	1
	3 Rationalised lead (lead strips bonded to face of single glass sheet)	Appearance inferior to traditional. Life related to durability of adhesive.	1
Technical Requirements Traditional: Water tight malleable metal framework to connect small panes into single window opening. Ideally long term durability to be as good as glass. Rationalised: Decorative effect to achieve similar appearance to traditional by adhesive bonding metal strip onto glass sheet. Adhesive must have good durability.	4 Rationalised lead (lead strips sealed in double glazing)	Appearance inferior to traditional. Adhesive protected from weathering.	1
Decay and Degradation Factors Natural weathering. Frost action. Wild life. Cleaning and cleaning solutions.			
Guidance Notes With this application a particular aesthetic effect is desired. In modern buildings the need for such an aesthetic effect is questionable, as it is preferable to avoid all the environmental and health impacts in the non-essential use of lead. For historic and other buildings of character it is suggested that copper be considered in preference to lead when complete replacement of the traditional came type is necessary, although this is not a standard alternative. Rubbing down or abrasion of existing lead cames for decoration and cleaning purposes should be avoided. Consideration could also be given to sealing the lead by means of varnish or paint. Abrasion should not be used to remove existing paint on lead cames. Secondary windows would assist in preventing children coming into contact with the cames. With modern buildings, if the leaded light effect must be provided, reproduction or facsimile appearance can be achieved by alternative 4 in preference to 3, as 4 prevents access to the lead. Traditional systems are marginally easier to recycle.			

Health Comment	Rank	Environmental Issues	Rank	Cost Rank
Dust generated by cleaning and maintenance is a potential contributor to lead uptake in young persons.	2/2	High embodied energy. Pollution from manufacturing. High recycled content. Assume material recycled at end of life.	2/0/0/0	118
No significant risk foreseen to occupants.	0/0	High embodied energy. Pollution from manufacturing. High recycled content. Assume material recycled at end of life.	2/0/0/0	272
No significant risk foreseen to occupants if applied outdoors. If applied indoors, rank as for traditional leaded lights.	0/0	High embodied energy. Pollution from manufacturing. Less material used than (1). Volatile organic compounds from adhesives. Very difficult to recycle.	2/0/0/1	100
No significant risk foreseen to occupants.	0/0	High embodied energy. Pollution from manufacturing. Less material used than (1). Volatile organic compounds from adhesives. Very difficult to recycle.	2/0/0/1	163

Application 3.10
GLAZED ROOF AND WALL STRUCTURAL SYSTEMS

Typical Situation

'O' ring gasket   cover clip

double glazing

ALUMINIUM/PVC SECTION

wall

double glazed units in sealant

bead

TIMBER SECTION

SECTION–GLAZED LEAN-TO

Technical Requirements

Method of constructing glazed areas giving minimum interruption to glazing, i.e. minimum size of glazing bar. Bars to be corrosion free, with adequate strength characteristics and ability to provide sealed cladding to resist air and rain penetration. Tidy visual appearance and convenient fixing required. Good overall thermal performance and minimum cold bridging.

Decay and Degradation Factors

Abrasion (sanding) if decoration required. Natural weathering. Frost action. Wild life. Corrosion. Aggressive water run-off from cementitious materials and metal flashings. Cleaning and cleaning solutions.

Guidance Notes

Appearance is an important criteria. Options 1 and 2 have the smallest cross-section and therefore the minimal visual intrusion, which might also be significant where light transmission is important. Options 3 and 5 are expensive cladding systems, incorporating a thermal break in the section and although not normally used in housing offer very good technical performance. Options 1-5 include gasket systems which provide good weather-proofing. Timber offers lower thermal bridging compared with normal (thermally unbroken) metal sections. Use of tropical hardwood should be discontinued – see Chapter 3.

Softwood and temperate hardwood offer lower environmental impact. From a health perspective, glazing poses a safety risk. Laminated or tempered safety glass should be used in lower panels, roof glazing and in doors.

Structural systems should be considered alongside glazing options.

Use of these systems to form conservatories can make a useful contribution to the thermal performance of houses.

However subsequent heating of such spaces for winter occupation as normal living spaces can seriously undermine the contribution of solar energy during the other periods of the year.

Alternatives	Technical Comment	Rank
1 Steel patent glazing systems with PVC wings and sheathing and neoprene seals	Development of traditional patent glazing systems. Durability dependent on life of gasket seals.	1
2 Aluminium patent glazing systems and wings with neoprene seals	Development of traditional patent glazing systems. Durability dependent on life of gasket seals.	1
3 Plastic coated steel curtain walling system	Complex propriety systems. Available with thermal break. Durability related to life of gasket seals.	1
4 Unplasticised Polyvinyl Chloride (PVCu)	Complex propriety systems. Available with thermal break. Durability related to life of gasket seals.	1
5 Aluminium curtain walling system	Complex propriety systems. Available with thermal break. Durability related to life of gasket seals.	1
6 Timber		
(a) Softwood	(a) Can be a durable solution if vacuum pregnated with preservative – if not, ranking 4. Requires regular maintenance.	2
(b) temperate hardwood	(b) Can be durable without other treatments.	2
(c) tropical hardwood	(c) Can be durable without other treatments.	2

Health Comment	Rank	Environmental Issues	Rank	Cost Rank
No significant risk foreseen to occupants.	0/0	Steel could be recycled but finishes cause problems. Problems from PVC sheathing as 4. Cold bridging – energy loss.	2/0/2/2	Refer to specialist manufacturers/ suppliers
No significant risk foreseen to occupants.	0/0	High embodied energy. Pollution from manufacture. High recycled content. Coatings may cause problems for recycling. Cold bridging – energy loss.	2/0/2/1	
No significant risk foreseen to occupants.	0/0	Steel could be recycled but finishes cause problems. Problems from PVC coating as 4.	2/0/1/1	
No significant risk foreseen to occupants.	0/0	Problems from pollution during manufacture. Problems at disposal. Galvanised/reinforcement steel. Sections similar to 1 above.	2/0/1/2	
No significant risk foreseen to occupants.	0/0	High embodied energy. Pollution from manufacture. High recycled content. Coatings may cause problems for recycling.	2/0/1/2	
No significant risk foreseen to occupants. (No substantial sanding foreseen)	0/0	Available from sustainable sources. Life expectancy related to maintenance and preservative treatments – see applications Paint and Wood Preservatives, 7.7, 7.9 and 7.10. Disposal by combustion causes atmospheric pollution.	1/0/1/0	
No significant risk foreseen to occupants. (No substantial sanding foreseen)	0/0	Availability from sustainable sources is limited. Upstream score 1 if from sustainable managed sources.	2/0/0/0	
No significant risk foreseen to occupants. (No substantial sanding foreseen)	0/0	Large proportion of tropical timber is still sourced from virgin forest. Upstream scores 1 if from sustainable managed sources.	3/0/0/0	

Application 4.1 INTERNAL PARTITIONS	Alternatives	Technical Comment	Rank
Typical Situation PLAN SECTIONS	1 Brick	Heavy, wet construction. Usually used to give fair faced 'brick' wall. Good sound insulation and fire properties.	1
	2 Block (concrete plastered)	Medium weight if 'lightweight' blocks used. Wet construction. Good sound insulation and fire properties.	1
	3 Softwood timber frame with facing options:	Traditional stud partitioning.	
	(a) plaster-board	(a) Good fire and sound properties possible. Requires joint treatment or skim coat of plaster. Susceptible to moisture damage.	1
Technical Requirements Self-supporting system for subdividing space. Ability to withstand lateral loading from shelving and other wall mounting may be required to carry dead and live loads from floors and roofs. Route for services (pipes and wiring). Ease of erection. Sound insulation may be important. Appearance. Fire resistance may be required. Ease of decoration and cleaning.	(b) calcium silicate boards	(b) Very good fire properties. Rigid material difficult to nail.	1
Decay and Degradation Factors Movements, insect and fungal attack. Fire. Wear and tear. Moisture from bathrooms, kitchen and plumbing leaks. Decoration, cleaning and cleaning solutions.	4 Galvanised frame with facing options:		
	(a) plaster-board	(a) Good fire and sound properties possible. Requires joint treatment or stain coat of plaster. Susceptible to moisture damage.	1
Guidance Notes Technical ranking is made on the basis of the appropriateness for the normal application and loadings, eg. it is less appropriate to use heavyweight blocks, supported on a timber floor. Stud partitions normally give adequate sound insulation between rooms in a dwelling, but would be inadequate between dwellings. Fire requirements may mean one hour performance in certain areas of flats. Options 1, 2 and 6 are normally used on ground floors or supported on concrete suspended floors, eg. as in flats. Options 1 and 6 are primarily used for decorative effects, although flettons plastered, form a common option to blockwork in the south of England. Supporting walls between dwellings require fire resistance and excellent sound insulation properties to comply with the building regulations. From an environmental perspective timber stud with plasterboard has a slightly lower impact over the other options, assuming the timber is recycled at end of life.	(b) calcium silicate boards	(b) Very good fire properties. Rigid material difficult to nail.	1
	5 Plasterboard systems (eg. paramount)	Prefabricated system – thinner therefore footprint on plan is less than other options – good fire properties.	2
	6 Glass blocks	Means of providing obscured glazed partition. Heavy. Systems using special fixing available avoids wet mortar construction.	2

Health Comment	Rank	Environmental Issues	Rank	Cost Rank
No significant risk foreseen to occupants.	0/0	Larger volume of material. Recycling problematic due to gypsum plaster contamination.	2/1/0/2	164
No significant risk foreseen to occupants.	0/0	Larger volume of material. Recycling problematic due to gypsum plaster contamination.	2/1/0/2	100
(a) No significant risk foreseen to occupants.	0/0	(a) Impacts due to plasterbord. Timber could be reused.	1/0/1/1	144
(b) No significant risk foreseen to occupants if undisturbed. Fibres may be released by abrasion, machining, maintenance or cleaning, or by the action of 'aggressive' water. Hazard to maintenance workers and DIY occupants depends upon the nature and quantity of the fibre.	0/0	(b) Impacts due to calcium silicate board facing. Timber could be reused.	2/0/0/1	188
No significant risk foreseen to occupants from concealed metal frame.				
(a) No significant risk foreseen to occupants.	0/0	(a) Impacts due to plasterboard and frame. Galvanised steel frame could be reused.	2/0/1/1	142
(b) No significant risk foreseen to occupants if undisturbed. Fibres may be released by abrasion, machining, maintenance or cleaning, or by the action of `aggressive' water. Hazard to maintenance workers and DIY occupants depends upon the nature and quantity of the fibre	0/0	(b) Impacts due to calcium silicate facing and frame. Galvanised steel frame could be reused.	2/0/0/1	185
No significant risk foreseen to occupants	0/0	Difficult to recycle.	1/0/1/2	190
No significant risk foreseen to occupants	0/0	Small quantity used for features.	2/0/0/0	2533

Application 4.2
CEILING AND WALL LININGS

Typical Situation

floor or roof joists

ceiling

partition framing

partition lining

Technical Requirements
Sheet material with adequate strength to span over framing members 400 – 600mm centres. Sound insulation properties may be important. Lining may be required to contribute to the fire resistance of the floor or wall. Non-combustibility may be a requirement but good fire properties (i.e. resistance to ignition and flame spread) an advantage. Fire properties may be modified by decoration. Water and rot resistance and advantage.

Decay and Degradation Factors
Abrasion due to sanding, scraping to remove wallpaper and other decoration. Drilling and cutting for fixing fittings and plumbing and electrical alterations. Impact. Normal wear and tear.

Guidance Notes
The major factor in selection is the required appearance, ie. painted plasterboard, timber or melamine finish. If a timber finish is required veneers use a lower volume of timber.

From a technical standpoint non-combustible linings may be required. Plasterboard ceilings are usually utilized to provide the major element in the ½ hour fire resistance required for the timber first floor or timber frame of 2-storey house construction. The material in two or more layers and/or combined with alternatives 7 and 8 can be used to provide the I hour (+) fire resistance for timber floors over integral garages or in flats.

Water resistant plasterboard performs better adjacent to showers, baths, or in kitchens. Ease of cutting and fixing (ie. drilling, nailing) is also a factor and this is a major reason for using plasterboard. Health concerns have been expressed over off-gassing from the formaldehyde adhesives in chipboard and to a lesser extent from plywood and blockboard. This can be minimised by selection of lower formaldehyde content adhesives.

Formaldehyde might become a cause for concern if two or more of the following are expected:

a. Large areas of fairfaced chipboard (ie. wall, ceiling and floor linings)

b. Urea-formaldehyde foam insulation was used in the cavity wall.

c. Low ventilation rates are expected.

A combination of these factors might cause irritation and/or reaction from sensitive individuals.

Timber boarding and timber based boards may pose a slight risk when treated with preservatives – see Application 7.7.

In environmental terms softwood and hardboard offer the lowest impact. Plasterboard is similar but it and hardboard are more difficult to recycle. Current construction practice often results in high wastage of plasterboard. Use of tropical hardwood should be discontinued. See also Application 4.3, Internal Wall Finishes.

Alternatives	Technical Comment	Rank
I Plasterboard	Used as dry liner decorated directly or with 3-5mm skim coat of plaster applied. Good fire properties; frequently used as fire protection lining.	I
2 Timber (tongue and groove boarding) (a) softwood (b) temperate hardwood (c) tropical hardwood	(a) Selected mainly for appearance. (b) Ditto (c) Ditto	2
3 Chipboard	Usually faced with plastic laminate, but could be used fair faced.	2
4 Blockboard	Usually faced with plastic laminate, but could be used fair faced.	2
5 Plywood	Can be used with plastic laminate facing but often used for natural appearance. V-grooved type available as cheaper sustitute for T & G boarding. Thinner sheeting requires closer spaced supports.	2
6 Hardboard	Can be used with range of surface finishes. Thinner sheeting requires closer spaced supports. May need conditioning.	4
7 Calcium silicate based board	Can be plastered if required. Good fire properties, some varieties non-combustible. Suitable for fire protection lining (especially as replacement for asbestos insulation board).	I
8 Glass reinforced cement board	Non-combustible. Suitable for fire protection lining (especially as a replacement for asbestos insulation board).	I
9 Asbestos Insulation Board	No longer available.	-

Health Comment	Rank	Environmental Issues	Rank	Cost Rank
No significant risk foreseen to occupants.	0/0	Moderate embodied energy. Impacts from gypsum extraction/conversion. Very difficult to recycle, if nailed and is plastered.	1/0/1/1	161+
No significant risk foreseen to occupants. Exposure to substantial amounts of fine wood dust is associated with adverse effects on breathing and with cancer, but would only be of concern in DIY sanding.	0/0	Disposal by combustion causes atmospheric pollution. (a) Available from sustainable sources. (b) Availability from sustainable sources is limited. Upstream score (1) if from sustainable managed sources. (c) Large proportion of tropical timber is still sourced from virgin forest. Upstream score (1) if from sustainable managed sources.	1/0/0/0 2/0/0/0 3/0/0/0	324+ 1434++ 1285++
No significant risk foreseen to occupants. (If poor ventilation allows volatile agents released from resin binders to build up, their acute irritant and long term effects and risk may be elevated to 0/3.	0/0	Resin adhesive. Upstream score (1) if fair faced.	2/0/1/1	110+
No significant risk foreseen to occupants. Exposure to substantial amounts of fine wood dust is associated with adverse effects on breathing and with cancer, but would only be of concern in DIY sanding.	0/0	Plywood facings usually contain tropical hardwood. Resin adhesive.	2/0/0/1	399+
No significant risk foreseen to occupants. Exposure to substantial amounts of fine wood dust is associated with adverse effects on breathing and with cancer, but would only be of concern in DIY sanding.	0/0	Plywood facings usually contain tropical hardwood. Resin adhesive.	2/0/0/1	132◆
No significant risk foreseen to occupants.	0/0	Assume softwood fibre from sustainable sources.	1/0/0/0	100◆
No significant risk foreseen to occupants if undisturbed. Fibres may be released by abrasion, machining, maintenance or cleaning or by the action of 'aggressive' water. Hazard to maintenance workers and DIY occupants depends on the nature and quantity of fibre.	0/0	Moderate embodied energy silicate from quarrying.	2/0/0/1	399❏
No significant risk foreseen to occupants, provided the diameters of the fibres have a low probability of being inhaled. Subsequent abrasion of the product will present no risk to maintenance workers or DIY occupants other than skin irritation.	0/0	Moderate embodied energy.	2/0/0/1	N/A
Fibre release on ageing, cleaning, maintenance and disposal presents a potential risk to maintenance workers, DIY occupants, construction and demolition operatives.	3/3	Asbestos is a hazardous waste and needs to be disposed of properly. Asbestos in existing buildings needs to be identified, recorded, assessed for comparative risks and removed or perhaps treated or encased and managed as appropriate. The (-) classification refers to lack of availability/no longer used.	-/-/-/3	N/A

◆ 4mm thick
❏ 6mm thick
+ 12mm thick
++ 16mm thick

Application 4.3 INTERNAL WALL FINISHES	Alternatives	Technical Comment	Rank
Typical Situation Decorative finishes to walls.	1 Plaster (gypsum)	In situ wet process. Drying time required. Conceals sub strata irregularities. Good fire and sound properties. Can provide high quality finish depending upon workmanship.	1
	2 Plaster (lime)	Similar to 1 above, but longer drying time required. Usually used in restoration work in historic buildings.	1
	3 Ceramic tiles	Easy clean surface. Useful in wet areas.	1
	4 Fair faced brick and block (undecorated)	Needs sealing with silicone or emulsion paint in order to control staining – see non-load bearing partitions.	2
Technical Requirements Appearance. Ability to mask background imperfections. Ease of cleaning. Resistance to cleaning solutions. Fire resistance/low flame spread my be required. Base for paint and wall paper finishes. May contribute to sound performance of wall. Ease of fixing or application.	5 Cork tiles	Improves sound absorption characteristics of room. Vulnerable in damp conditions. Needs sealing to prevent staining over time.	2

Decay and Degradation Factors

Impact, U.V., rot and mould, cleaning and cleaning solutions, structural movement. Insects and vermin. Redecoration. Fire. Dampness.

Guidance Notes

The main selection factors are concerned with the required appearance in relation to the function, eg. Ceramic tiles in bathroom for ease of cleaning, cork tiles for acoustic damping etc.

The choice between lime and gypsum plasters relates to speed of set, impact resistance, nature of the background and ability to tolerate minor movement in walls. Both can give good quality finish with appropriate workmanship, although lime plaster offers slightly lower environmental impact.

Fair faced masonry walls require more care when units are laid with a slight time penalty. In principle, as no additional finishing material is used, this option offers lower environmental impact. However, these, as well as 2 and 5, may require more maintenance, cleaning, sealing, etc, to maintain the original appearance. Cleaning may create separate negative environmental impacts depending upon solutions used.

Other health and environmental effects may accrue from the adhesives utilised for options 3 and 5 – see Application 7.4.

Health Comment	Rank	Environmental Issues	Rank	Cost Rank
No significant risk foreseen to occupants.	0/0	Moderate embodied energy. Impacts from gypsum extraction/conversion.	2/0/0/1	175
No significant risk foreseen to occupants.	0/0	Impacts from extraction of lime and sand	1/0/0/0	184
No significant risk foreseen to occupants.	0/0	High embodied energy and atmospheric emissions.	2/0/0/0	633
No significant risk foreseen to occupants.	0/0	No finish – impacts of bricks and blocks see application 7.3. More frequent cleaning required.	0/0/1/0	100
No significant risk foreseen to occupants. Resurfacing with polymeric sealants by DIY may present a respiratory hazard.	0/0	Natural material from sustainable source. Adhesives usually contain Volatile Organic Compounds.	1/0/0/0	364

Application 4.4 ARCHITRAVES AND SKIRTINGS	Alternatives	Technical Comment	Rank
Typical Situation	1 Timber		
	(a) Softwood	(a) Adequate solution, usually decorated.	1
	(b) Temperate hardwood	(b) Selected for appearance. More durable in damp areas.	1
	(c) Tropical hardwood	(c) Selected for appearance. More durable in damp areas.	1
	2 Unplasticized polyvinyl chloride	Useful in wet areas. Usually fixed with adhesive.	1
Technical Requirements Cover strip to conceal joint between structure and other features such as doors and window frames. May need to cope with movements of substrate. Good appearance. Ease of cleaning and decoration. Ease of fitting.	3 Terrazzo (skirtings only)	Hard wearing, durable, easy to clean – applied to match terrazzo floor finish.	1
Decay and Degradation Factors Impact. Movement. Fungal and insect attack. Decoration. Cleaning and cleaning solutions.	4 Galvanised steel plaster stops (and sealants)	Avoids need for architraving and skirtings. Good workmanship in fitting and plastering required to achieve good finish.	2
Guidance Notes The main selection factors are concerned with the required appearance in relation to the function, eg. PVC, terrazzo and quarry tiles, in wet areas.	5 Quarry tiles (skirtings only)	Similar performance to terrazzo with quarry tiled floors.	1
In the majority of internal application in dwellings softwood is adequate and offers the lowest environmental impact. It appears to be possible to avoid even this by detailing the junctions without the use of architraves and skirting but in reality this requires use of galvanised steel plaster stops with good workmanship. Beadwork/sealants may also be required (see application 7.4 and 7.5). Together this is likely to result in a greater impact. Use of tropical hardwoods should be avoided – see Chapter 3.	6 Medium Density Fibreboard	Useful where complex mouldings required.	1

Health Comment	Rank	Environmental Issues	Rank	Cost Rank
(a) – (c) No significant risk foreseen to occupants. Exposure to substantial amounts of fine wood dust is associated with adverse effects on breathing and with cancer, but would only be of concern in DIY sanding.	0/0	Disposal by combustion causes atmospheric pollution. (a) Available from sustainable sources. (b) Availability from sustainable sources is limited. Upstream score (1) if from sustainable managed sources. (c) Large proportion of tropical timber is still sourced from virgin forest. Upstream score (1) if from sustainably managed sources.	1/0/0/0 2/0/0/0 3/0/0/0	178 304 273
No significant risk foreseen to occupants.	0/0	Problems from pollution during manufacture, but situation is improving. Problems at disposal – modest volumes.	2/0/0/1	195
No significant risk foreseen to occupants.	0/0	Moderate embodied energy. Very durable. Impact from aggregate extraction.	2/0/0/0	3008
No significant risk foreseen to occupants.	0/0	Small quantities of steel. Pollution from galvanised coating. Very difficult to recycle. Pollution from sealant manufacture.	2/0/1/1	100
No significant risk foreseen to occupants.	0/0	High embodied energy and atmospheric emissions.	2/0/0/0	480
No significant risk foreseen to occupants.	0/0	Upstream score (1) if fibre sourced from forest trimmings and waste. Resin binder.	2/0/0/0	143

Application 4.5 FLOORBOARDING	Alternatives	Technical Comment	Rank
Typical Situation floorboarding floorcovering or direct varnish finish floor joists ceiling	1 Softwood timber tongue and groove boarding (T & G)	Traditional flooring. Permits relatively easy access to floor voids for alteration and maintenance. Sound and fire performance of floor construction using T & G superior to plane edge boarding.	1
	2 Temperate hardwood. timber tongue and groove boarding	Selected for appearance.	1
	3 Tropical hardwood	Selected for appearance.	1
Technical Requirements Smooth floor surface. Nailable material to span normal joist spacing of 400-600mm centres. Ease of cutting essential. Boarding contributes to fire and sound performance of total floor construction. Good fire properties (i.e. resistant to ignition from small fire sources and resistance to flame spread) an advantage.	4 Chipboard (T & G edged)	Purpose made flooring sheets available. Severe dampness can cause loss of strength. Water resistant grade should be used in kitchen and bathroom areas. Access to floor void can be more difficult.	2
Decay and Degradation Factors Abrasion due to foot traffic will take place unless protected by floor coverings. Fire. Sanding down for re-varnishing where applicable.	5 Blockboard	Rarely used for this application but could prove satisfactory. Not available as proprietary product with T & G edge.	1
Guidance Notes Options 1 and 3 are the most common selection for domestic floors. From a technical viewpoint selection is based mainly upon cost/convenience factors particularly in relation to access to the plumbing and wiring systems located in the floor void. Health concerns have been expressed over off-gassing of the formaldehyde from the resin adhesives in chipboard and to a lesser extent from plywood and blockboard. This can be minimised by selection of products utilising lower formaldehyde content adhesives. However, this is still a concern where large areas of material is used with very low rates of ventilation (see also application 4.2). Maintenance requiring sanding-down may generate risks from wood dust. D.I.Y. operators should wear protective masks. Hardwood flooring is available in thin sections for overlaying timbers or solid floors in a similar manner to wood blocks. This uses less material and is prefered environmentally. Use of tropical hardwoods should be discontinued, see chapter 3.	6 Plywood	Greater bending strength allows greater span between joist centres. Otherwise as 5.	

Health Comment	Rank	Environmental Issues	Rank	Cost Rank
No significant risk foreseen to occupants. Exposure to substantial amounts of fine wood dust is associated with adverse effects on breathing and with cancer, but would only be of concern in DIY sanding.	0/0	Available from sustainable sources. Disposal by combustion causes atmospheric pollution.	1/0/0/0	239
No significant risk foreseen to occupants. Exposure to substantial amounts of fine wood dust is associated with adverse effects on breathing and with cancer, but would only be of concern in DIY sanding.	0/0	Availability from sustainable sources is limited. Upstream score (1) if from sustainable managed sources.	2/0/0/0	1164+
No significant risk foreseen to occupants. Exposure to substantial amounts of fine wood dust is associated with adverse effects on breathing and with cancer, but would only be of concern in DIY sanding.	0/0	Large proportion of tropical timber is still sourced from virgin forest. Upstream score (1) if from sustainable managed sources.	3/0/0/0	1043++
No significant risk foreseen to occupants. If poor ventilation leads to a build up of acute respiratory irritant and carcinogenic vapours derived from binding resin, the risk may be elevated to unacceptable 0/3.	0/0	Assume softwood fibre from sustainable sources. Resin adhesive.	2/0/1/1	100
No significant risk foreseen to occupants. If poor ventilation leads to a build up of acute respiratory irritant and carcinogenic vapours derived from binding resin, the risk may be elevated to unacceptable 0/3.	0/0	Plywood facings usually contain tropical hardwood. Resin adhesive.	2/0/0/1	269
No significant risk foreseen to occupants. If poor ventilation leads to a build up of acute respiratory irritant and carcinogenic vapours derived from binding resin, the risk may be elevated to unacceptable 0/3.	0/0	Plywood facings usually contain tropical hardwood. Resin adhesive.	2/0/0/1	242

+ Oak
++ Mahogany

Application 4.6 FLOOR TILE AND SHEET	Alternatives	Technical Comment	Rank
Typical Situation CONCRETE FLOOR tiles / adhesive / floor scread concrete / hardcore base adhesive tiles / hardboard / floorboards / joists TIMBER SUSPENDED FLOOR	1 Flexible PVC (poly-vinyl)-mineral fibre	Good fire properties. Smooth level surface required.	1
	2 Flexible PVC (Vinyl)	More easily damaged by cigarette burns than alternatives. Smooth level surface required.	1
	3 Thermo-plastic (asphalt tiles)	Restrictive colour range compared with alternatives. Smooth level surface required.	2
	4 Ceramic	Very hard wearing. Good fire properties.	1
Technical Requirements Smooth but slip resistant easily cleaned surface of acceptable appearance. Hardwearing surface resistant to damage from heel load/castor, etc. Resistant to water and attack by cleaning solutions, fats, cooking oils, etc. Dimensional stability vital. Good fire properties (i.e. resistant to ignition from small fire sources and resistant to flame spread) an advantage.	5 Wood blocks and strip (a) temperate hardwood (b) tropical hardwood	(a) Susceptible to moisture movement if not properly sealed. (b) Ditto	1 1
Decay and Degradation Factors Ordinary wear and tear, foot traffic and spillage. Potential attack from cleaning powders/solutions. Possible removal by DIY owner. Abrasion from furniture movement. Fire.			
Guidance Notes Technical selection depends upon required wear and moisture resistance, as well as ease of cleaning, which in turn is related to location in the dwelling (e.g. hall, lounge, bathroom, etc) and also upon particular manufacturer's formulations.	6 Cork	Susceptible to moisture movement if not properly sealed and more easily damaged by impact. Smooth level surface required.	3
From a health perspective the main concerns centre over fibre release from option 1 and wood dust from maintenance of option 5. D.I.Y. operators should wear dust masks.	7 Terrazzo	Very hard wearing. Good fire properties.	1
Existing vinyl and thermoplastic asbestos floors present a significant risk of fibre release especially at replacement. This material may still be present in buildings and precautions should be taken at removal. It must be disposed of properly in accordance with waste regulations.	8 Quarry tiles	Very hard wearing. Good fire properties.	1
Linoleum and natural stone offer low environmental impact. Cork is a 'renewable' material, is 'warm' and provides acoustic damping, but requires higher maintenance and has poorer life expectancy. Use of tropical hardwoods should be avoided – see Chapter 3.	9 Linoleum	Smooth level surface required.	2
	10 Stone	Very hard wearing. Good fire properties.	1
Woodblocks can be recycled. Natural oil based finishes are to be prefered over solvent based varnishes or wood blocks and strip.	11 Vinyl and thermo-plastic asbestos fibre	No longer available but may be present in existing buildings.	-

Health Comment	Rank	Environmental Issues	Rank	Cost Rank
No significant risk foreseen to occupants so long as the PVC matrix is intact. If a substantial amount of inhalable material fibre is released by attrition, risk may be elevated.	0/1	Problems from pollution during manufacture but situation is improving. Adhesives usually contain Volatile Organic Compounds.	2/0/1/1	238
No significant risk foreseen to occupants.	0/0	Problems from pollution during manufacture but situation is improving. Adhesives usually contain Volatile Organic Compounds.	2/0/1/1	127
No significant risk foreseen to occupants.	0/0	Problems from pollution during manufacture. Adhesives usually contain Volatile Organic Compounds.	2/0/1/1	100
No significant risk foreseen to occupants.	0/0	High embodied energy and atmospheric emissions.	2/0/0/0	334
No significant risk foreseen to occupants. Exposure to substantial amounts of fine wood dust is associated with adverse effects on breathing and with cancer, but would only be of concern in DIY sanding.	0/0	(a) Availability from sustainable sources is limited. Upstream score 1 if from sustainable managed sources. Volatile Organic Compounds in some varnishes. (b) Large proportion of tropical wood is still sourced from virgin forest. Upstream score 1 if from sustainable managed sources. Volatile Organic Compounds in some varnishes.	2/0/0/0 3/0/0/0	671+ 617++
No significant risk foreseen to occupants. Resurfacing with polymeric sealants by DIY may present a respiratory hazard.	0/0	Natural material from sustainable source. Volatile Organic Compounds in some varnishes and adhesives. Relatively poor durability.	1/0/2/0	206
No significant risk foreseen to occupants.	0/0	Moderate embodied energy. Impact from aggregate extraction.	2/0/0/0	377
No significant risk foreseen to occupants.	0/0	High embodied energy and atmospheric emissions.	2/0/0/0	293
No significant risk foreseen to occupants.	0/0	Mainly natural materials. Adhesives usually contain Volatile Organic Compounds.	1/0/0/0	192
No significant risk foreseen to occupants.	0/0	Quarrying and quarry waste.	1/0/0/0	950
The release of inhalable fibre by wear has led to the long term risk from this application being acknowledged.	1/3	Asbestos is a hazardous waste and needs to be disposed of properly. Asbestos in existing buildings needs to be identified, recorded, assessed for comparative risks and removed or perhaps treated or encased and managed as appropriate. The (-) classification refers to lack of availability/no longer used.	-/-/-/3	-

+ Oak
++ Mahogany

Application 4.7(a) **CARPETS AND CARPET TILES**	**Alternatives**	**Technical Comment**	**Rank**
Typical Situation NORMAL TUFTED CARPET — tufts pushed through — base fabric — jute backing FOAM BACKED TUFTED CARPET — tufts — base fabric — foam backing	**Carpet fibre/pile** 1 Wool	Traditional material. Reasonably durable and `springy'. Easiest to clean and keeps its appearance well.	1
Technical Requirements Wearing properties selected as appropriate to the use of the room. Sound absorption and insulation properties may be useful. Thermal properties a consideration. Resistant to decay in normal use. Easy to clean. Aesthetics important.	2 Nylon	Hardwearing alternative to wool. Often blended with wool. Static sometimes a problem. Uses stain guards.	1
	3 Polyester	Used minimally in carpet manufacture. Hardwearing alternative to wool. Often blended with wool. Static sometimes a problem. Uses stain guards.	1
Decay and Degradation Factors Dampness. Fungal, insect and vermin attack. Wear from foot traffic. Degradation from spillages and UV light, and discolouration from soiling.	4 Poly-propylene	Use is growing in popularity. Hardwearing alternative to wool. Often blended with wool. Uses stain guards. Resistant to staining including bleach.	1
Guidance Notes Wear performance depends on the complete system, pile type and density, backing and appropriate underlay where required. Plastics based fibres require chemical stainguards to provide adequate stain resistance. Some cheaper foam underlays give poor durability.	5 Acrylic	Used mainly as synthetic Axministers and Wiltons. Hard wearing.	1
Under warm, humid environmental conditions, carpets will become infested with house mites that are powerful allergens and a common cause of asthma. Ordinary vacuum cleaning is not effective in their removal. Certain individuals may suffer allergic reaction to off-gassing released from adhesives in newly laid carpets.	**Carpet backing and underlay** 6 Hessian (backing)	Traditional backing a. Woven – hardest wearing b. Secondary stock – not as hard wearing.	 1 2
Natural fibre and material offer the lowest environmental impact, eg. Option 1 for pile, options 6, 7, 9 and 11 for backing underlay.	7 Felt	Traditional underlay. Better than rubber (optons 9,10 and 11) but compresses over a period of time.	2
The question can be posed, from both a health and environmental perspective, whether carpets should be utilised at all.	8 Polyurethane foam	Needs to be stapled or glued at edge. Tends to disintegrate after time especially if exposed to moisture.	3

Health Comment	Rank	Environmental Issues	Rank	Cost Rank
No significant risk foreseen to occupants.	0/0	Natural renewable resource. Biodegradable – life expectancy (Cscore) reduces to (1) if heavy foot traffic. Some pollution from dye manufacture and sheep dipping. Downstream 1 if incinerated.	0/0/0/0	242+*
No significant risk foreseen to occupants.	0/0	Relatively high energy in manufacture. Higher levels of pollution from petrochemical production. Releases toxic fumes if incinerated.	2/0/0/1	100*
No significant risk foreseen to occupants.	0/0	Relatively high energy in manufacture. Pollution from petrochemical production. Problems of disposal, releases toxic fumes if incinerated.	1/0/0/1	179*
No significant risk foreseen to occupants.	0/0	Relatively high energy in manufacture. Pollution from petrochemical production. Problems of disposal, releases toxic fumes if incinerated.	1/0/0/1	268*
No significant risk foreseen to occupants.	0/0	Relatively high energy in manufacture. Pollution from petrochemical production. Problems of disposal, releases toxic fumes if incinerated.	1/0/0/1	303*
No significant risk foreseen to occupants.	0/0	Natural material. Disposal could rise to 1 if manufactured with synthetic materials.	0/0/0/0	N/A
No significant risk foreseen to occupants.	0/0	Felt can be manufactured from recycled fabric.	0/0/0/0	100*
No significant risk foreseen to occupants.	0/0	Relative high energy and pollution during manufacture. Disposal problematic.	2/0/1/1	N/A

* supply only
+ 20% nylon

Application 4.7(b) CARPETS AND CARPET TILES	Alternatives	Technical Comment	Rank
	Carpet backing and underlay 9 Latex	Used extensively for underlay. Natural latex rubber has a long life but when mixed with fillers (eg. chalk) life is reduced.	3
	10 Latex foam	Used extensively. Foam can become brittle over time.	3
	11 Felt and rubber composition	50% felt, 50% rubber. Rubber sourced from recycled tyres. Durable.	1

Health Comment	Rank	Environmental Issues	Rank	Cost Rank
No significant risk foreseen to occupants.	0/0	Natural rubber, although synthetic rubber is also used. Upstream grading assumes natural rubber otherwise grade will rise to 1. Downstream 1 if incinerated at end of life.	0/0/0/0	N/A
No significant risk foreseen to occupants.	0/0	Natural rubber, although synthetic rubber is also used. Upstream grading assumes natural rubber otherwise grade will rise to 1. Uses blowing agents during manufacture. Environmental impacts if disposed by incineration not known.	0/0/0/1	N/A
No significant risk foreseen to occupants.	0/0	Uses natural and recycled products. Downstream 1 if incinerated.	0/0/0/0	144*

* supply only

Application 4.8
EXTERNAL WALL FINISHES

Typical Situation

horizontal battens   vertical battens   vertical battens or clip rails   joints raked-out for key

slates

render

SLATES/ TILES   TIMBER BOARDING   PLASTIC BOARDING   RENDERING

Technical Requirements
Effective means of providing water-proof layer to the external face of walls.

Decay and Degradation Factors
Normal weathering, frost, high winds, ultra-violet, chemical attack – cleaning, acid rain, incompatible materials. Impact, graffiti, fungal attack – particularly to timber support battens.

Guidance Notes
Selection relates to a balance between durability and appearance factors. The main choice of appearance is between tiled, planked, render or sheeted material, each of which is distinctive. Durability is very dependent upon the fixing systems employed and upon workmanship. Softwoods require preservative treatments plus painting or varnishing to provide adequate durability.

In terms of a planked appearance, timber boards are marginally easier to cut and fix. For sheet material jointing is an important factor, as this is often achieved using timber battens.

Rendering is a wet process which can be slow and messy.

There is little to choose between tiles or slates from an environmental perspective, although slate has slightly better durability.

For a planked appearance softwood has slightly lower environmental impact and for rendering lime is marginally better from this point of view.

Use of tropical hardwoods should be discontinued, see chapter 3.

Alternatives	Technical Comment	Rank
1 Concrete tiles on timber battens	Interlocking concrete slates or tiles lighter than plain slates or tiles, which require to be laid double lap. Both types give good durability and are non-combustible.	1
2 Clay tiles on timber battens	Interlocking concrete slates or tiles lighter than plain slates or tiles, which require to be laid double lap. Both types give good durability and are non-combustible.	1
3 Slate on timber battens	Proven, very durable material. Non-combustible.	1
4 Timber boards on timber battens	Various board profiles. Easy to cut and fit.	
(a) Softwoods	(a) Decoration required for protection.	2
(b) Temporate hardwoods	(b) Can be durable without other treatments.	1
(c) Tropical hardwoods	(c) Can be durable without other treatments.	1
5 Unplasticised Polyvinyl Chloride (PVCu) – cellular planks battens or clip fix systems	Requires careful setting out as impracticable to cut planks along the length. Otherwise easy to fit. Gradually degraded by UV light.	1
6 Polyvinyl Chloride (PVC) planks on battens or clip fix systems	Requires careful setting out as impracticable to cut planks along the length. Otherwise easy to fit. Gradually degraded by UV light.	1
7 Calcium silicate boards or sheets on battens	Reasonable workability characteristics, but difficult to nail. Good fire properties. Board quality determines performance – select accordingly.	1
8 Glass Reinforced Plastic (GRP) Planks	Requires careful setting out as impracticable to cut planks along the length. Otherwise easy to fit. Gradually degrade by UV light.	1
9 Cement render	Good durability provided attention is given to preparation of background and good standard of workmanship.	2
10 Lime mortar and render	Good durability provided attention is given to preparation of background and good standard of workmanship Longer drying time can cause problems in poor weather.	1

Health Comment	Rank	Environmental Issues	Rank	Cost Rank
No significant risk foreseen to occupants.	0/0	Moderate embodied energy. Potential for recycling.	1/0/1/2	902+
No significant risk foreseen to occupants.	0/0	Higher embodied energy and atmospheric emissions. A percentage can be reused.	2/0/1/1	863
No significant risk foreseen to occupants.	0/0	Good life expectancy, capable of reuse many times. Despolation by quarry waste.	2/0/0/1	1138
(a) – (c) No significant risk foreseen to occupants. Exposure to substantial amounts of fine wood dust is associated with adverse effects on breathing and with cancer, but would only be of concern in DIY sanding.	0/0	Disposal by combustion causes atmospheric pollution. (a) Available from sustainable sources. Life expectancy related to maintenance and preservative treatments – see applications Paint and Wood Preservatives, 7.7, 7.9 and 7.10 (b) Availability from sustainable sources is limited. Upstream score 1 if from sustainably managed sources. (c) Large proportion of tropical timber is still sourced from virgin forest. Upstream scores 1 if from sustainably managed sources.	1/0/1/0 2/0/0/0 3/0/0/0	388 1470 1357
No significant risk foreseen to occupants.	0/0	Problems from pollution during manufacture but situation is improving. Problems at disposal.	2/0/1/2	442
No significant risk foreseen to occupants.	0/0	Problems from pollution during manufacture but situation is improving. Problems at disposal.	2/0/1/2	473
No significant risk foreseen to occupants.	0/0	Moderate embodied energy. Impacts from quarrying.	2/0/1/1	226
No significant risk foreseen to occupants.	0/0	Resin binders. Pollution during manufacture and disposal.	2/0/1/2	N/A
No significant risk foreseen to occupants.	0/0	Pollution from cement manufacture. Impacts from quarrying for sand. Problem of separation from bricks at demolition may inhibit brick recycling	1/0/0/1	100
No significant risk foreseen to occupants.	0/0	Durability suspect unless well maintained using limewash. Relatively easy to separate from bricks at demolition facilitates brick recycling.	0/0/1/0	N/A

+ concrete tiles

Application 5.1
ABOVE GROUND DRAINAGE

Typical Situation

Technical Requirements

Effective means of discharging sewage and grey water to underground system. Good durability and ability to withstand impact and movement. Fire resistance may be required in certain circumstances. Appearance and ease of fixing may be important.

Decay and Degradation Factors

Frost, birds and vermin, UV, impact, vandalism, movements, chemical and cleaning solutions.

Guidance Notes

Copper and lead were extensively used in these applications in the past, but have been replaced with the alternatives shown. These perform adequately for the purpose. Plastic systems are lighter and easier to cut and fit. Polypropylene and ABS offer cost advantages and lower environmental impact.

Manufacturers instructions should be followed to prevent sagging.

Alternatives	Technical Comment	Rank
1 Polypropylene (PP)	Used for soil, and waste pipes, and traps.	1
2 Acrylonitrile Butadiene Styrene (ABS)	Used for soil and waste pipes.	1
3 Unplasticised Polyvinyl Chloride (PVCu)	Used for soil and waste pipes.	1
4 Modified PVCu	Used for soil and waste pipes.	1
5 Polyethelene (PE)	Mainly for traps.	1
6 Cast Iron using sealed spigot and socket or PVCu joint fittings	Normally used for replacement work (especially rainwater goods, see Application 1.6).	1

Health Comment	Rank	Environmental Issues	Rank	Cost Rank
No significant risk foreseen to occupants.	0/0	Moderate embodied energy. Pollution from petrochemicals. Downstream score assumes some recycling.	1/0/0/1	112
No significant risk foreseen to occupants.	0/0	Moderate embodied energy. Pollution from petrochemicals. Downstream score assumes some recycling.	1/0/0/1	100
No significant risk foreseen to occupants.	0/0	Problems from pollution during manufacture but situation is improving. Pollution from petrochemicals. Problems at disposal.	2/0/0/2	184
No significant risk foreseen to occupants.	0/0	Problems from pollution during manufacture but situation is improving. Pollution from petrochemicals. Problems at disposal.	2/0/0/2	141
No significant risk foreseen to occupants.	0/0	Moderate embodied energy. Pollution from petrochemicals. Downstream score assumes some recycling.	1/0/0/1	N/A
No significant risk foreseen to occupants.	0/0	High embodied energy. Pollution from metal smelting. Downstream score assume high % recycled.	2/0/0/0	454

Application 5.2
UNDERGROUND DRAINAGE

	Alternatives	Technical Comment	Rank
Typical Situation 	**Pipes & Fittings*** 1 Vitrified clay pipes and fittings (spiget and socket)	Glass fibre yarn used in caulking. Slight disadvantage in time taken to make joints. Susceptible to excessive ground movements.	2
	2 Vitrified clay pipes with plastic fittings (see plastic pipes below, ie. fittings)	Less susceptible to ground movements than 1.	1
	3 Cast iron a. Spigot and socket b. PVCu joint fittings	Normally used for replacement work in historic buildings – rarely used in new construction.	1 1
Technical Requirements Durable, adequate strength to resist wheeled vehicle loads in certain locations. Ability to withstand movement. Chemical resistance. Ease of fixing and laying.	4 Concrete spigot and socket pipes	Mainly used in larger diameter main sewers.	1
Decay and Degradation Factors Frost, settlement, vermin, chemicals and contaminants in the ground and effluent carried by pipe. Loads from vehicles and surface.	5 High density polyethylene (HPDE)	Good durability in domestic applications. Pipe available in longer lengths with push fit connectors.	1
Guidance Notes The bedding and backfilling of pipes at laying is very significant to long term durability. Manufacturers' recommendations for the size of gravel bedding and depth should be followed. Plastic pipe, joint and manhole systems are flexible and will accommodate some ground movement. These systems are easier and quicker to cut and fit. Manholes located in roadways or that are very deep usually require options 9 or 10.	6 Polypropylene (PP)	Good durability in domestic applications. Pipe available in longer lengths with push fit connectors.	1
	7 PVCu	Good durability in domestic applicatons.	1
The downstream environmental impact is not assessed as pipework is difficult to recover and so is most likely to be abandoned in the ground. Also the scores ignore the impacts accruing due to the bedding aggregates – see Sheet 7.2.	**Manholes** 8 Un-plasticised Polyvinyl chloride (PVCu)	Good durability in domestic applications. Pipe available in longer lengths with push fit connectors.	1
For shallow manholes G.R.P. or other plastics will use less material than brick or concrete.	9 Glass reinforced plastic	Framed in various sizes. Easy connection to plastic system.	1
	10 Engineering brick	Established construction technique. Slow but flexible solution for non standard or repairs.	1
	11 Concrete	Manholes pre-cast in various sizes.	1

* fittings include joints, brackets etc

Health Comment	Rank	Environmental Issues	Rank	Cost Rank
No significant risk foreseen to occupants.	0/0	High embodied energy and atmospheric emissions.	2/0/0/-	195+
No significant risk foreseen to occupants.	0/0	High embodied energy and atmospheric emissions.	2/0/0/-	202+
No significant risk foreseen to occupants.	0/0	High embodied energy. Pollution from metal smelting.	2/0/0/-	1140+ 830+
No significant risk foreseen to occupants.	0/0	Moderate embodied energy. Pollution from cement manufacture.	1/0/0/-	304++
No significant risk foreseen to occupants.	0/0	Moderate embodied energy. Pollution from petrochemicals.	1/0/0/-	N/A
No significant risk foreseen to occupants.	0/0	Moderate embodied energy. Pollution from petrochemicals.	1/0/0/-	N/A
No significant risk foreseen to occupants.	0/0	Moderate embodied energy. Pollution from petrochemicals.	1/0/0/-	100
No significant risk foreseen to occupants.	0/0	Problems of pollution during manufacture but situation is improving. Pollution from petrochemicals.	2/0/0/-	N/A
No significant risk foreseen to occupants.	0/0	Resin binders – pollution during manufacture.	2/0/0/-	N/A
No significant risk foreseen to occupants.	0/0	High embodied energy. Atmospheric emissions. (Score based on bricks but also consider concrete/mortar.)	2/0/0/-	100*
No significant risk foreseen to occupants.	0/0	Moderate embodied energy. Pollution from cement manufacture.	1/0/0/-	128*

+ 150mm dia
++ 225mm dia
* 650 x 450 x 750 deep

Application 5.3 HOT AND COLD WATER PIPEWORK	**Alternatives**	**Technical Comment**	**Rank**
	1 Copper	Well-proved material.	1
	2 Stainless steel	More difficult to work than copper. Needs special flux for soft solder capillary fittings.	1
	3 Cross linked poly-ethylene	Uses compression fittings. Flexible and available in long lengths. Flexible pipe.	1
	4 Chlorinated PVCu	Used for pipes and fittings.	1
	5 Polybutylene	Used for pipes and fittings	1
	6 Poly-propylene	Used for pipes and fittings	1
	7 Polyvinyl-idiene Fluoride	Used for pipes and fittings.	1
	8 Polysulphene	Used for pipes and fittings.	1
	9 Lead	No longer available.	

Typical Situation

PLASTIC PUSH-FIT   METAL SOLDER   METAL COMPRESSION

Technical Requirements

Below Ground: Impervious durable pipe system to connect from water mains to consumer's stop cock within the dwelling without contamination of supply water.
Resistance to attack from ground water. Crushing by overburden particularly when drained. Normally laid below frost line.
Other Applications: Pipe systems of adequate strength and stability to resist temperature and pressure fluctuations and that are easy to install.

Decay and Degradation Factors

Below Ground: Attack from ground water which may contain a wide range of aggressive chemical compounds – especially when site redeveloped from tip or industrial usage. Possible frost action at access points. Movement. Settlement.
Other Applications: Impact, high temperatures, movements, frost, chemical solutions in water, U.V. light where exposed.

Guidance Notes

The industry has moved away from metal tube and fittings to plastics systems below ground, eg. water supply pipework, as well as for a range of other plumbing applications. The main advantage of these systems is that they are quicker and easier to use (although some fittings are more expensive). Plastic also offers some resistance to frost, but all pipework above ground needs to be insulated to protect from frost – see Sheet 2.8.

When jointing involves the use of solder, applications 1 and 2, the toxic potential of constituents of the alloy to contaminate potable water, needs to be taken into consideration. In the case of polymeric products, their health status is based on the assumption that they conform to standards requiring additives, that may leak out to be non-toxic.

Lead supply pipework including mains distribution pipes were common in older property. When major alterations or refurbishment is contemplated it is advisable and may prove convenient to replace lead pipework while work is in progress. Otherwise the advice of the local environmental health officer should be sought in respect of the local water quality to ascertain whether immediate action should be taken to replace existing lead pipe.

The environmental ranking assumes metals are recycled at end of life and does not consider impacts from metal (eg. brass) compression fittings which have slightly higher environmental impact.

Health Comment	Rank	Environmental Issues	Rank	Cost Rank
No significant risk foreseen to occupants.	0/0	High embodied energy. Pollution from manufacturing. High recycled content. Depletion of scarce resource.	2/0/0/0	305
No significant risk foreseen to occupants.	0/0	Score assumes UK sources with 100% recycled content.	1/0/0/0	420
No significant risk foreseen to occupants.	0/0	Moderate embodied energy. Pollution from petrochemicals. Downstream score assumes some recycling.	1/0/0/1	100
No significant risk foreseen to occupants.	0/0	Moderate embodied energy. Pollution from petrochemicals. Downstream score assumes some recycling.	2/0/0/1	195
No significant risk foreseen to occupants.	0/0	Moderate embodied energy. Pollution from petrochemicals. Downstream score assumes some recycling.	1/0/0/1	213
No significant risk foreseen to occupants.	0/0	Moderate embodied energy. Pollution from petrochemicals. Downstream score assumes some recycling.	1/0/0/1	458
No significant risk foreseen to occupants.	0/0	Moderate embodied energy. Pollution from petrochemicals. Downstream score assumes some recycling.	1/0/0/1	N/A
No significant risk foreseen to occupants.	0/0	Moderate embodied energy. Pollution from petrochemicals. Downstream score assumes some recycling.	1/0/0/1	N/A
Toxic	3/3	Lead pipes no longer specified, therefore no classification (-). Score 'O' downstream if material is recycled. This is the preferred disposal method for this metal.	-/-/-/3	N/A

Application 5.4
COLD WATER STORAGE TANKS

Typical Situation

insulation

cover

tank

bearers

roof timbers

ceiling

Technical Requirements

Container for water storage must provide flexibility/ability to cut for feed and outflow pipes/valves, etc. Cover essential. Self supporting over bearers an advantage. Material of tank should not contaminate water supply.

Decay and Degradation Factors

Danger might result from plumbing alterations where fibres are left in tank from cutting holes. Removal of lid for maintenance. Possible attack of material by aggressive waters causing release of particles and fibres. Build-up of sediment and corrosion may exacerbate this. Risk of electrolytic corrosion in metals.

Guidance Notes

With very acidic waters durability of galvanised steel tanks may be limited.

Plastic tanks are obviously susceptible to ignition and care is required during plumbing operations which might involve the use of hot work with a naked flame (soldering, brazing, etc).

Otherwise, the choice is dependent upon cost/convenience factors (i.e. shape/size of tank, position in roof, etc) although 'plastics' types, alternatives 2, 3 and 4 are now most common.

By this date it is likely that most existing asbestos cement tanks will have been replaced. If not, any existing tanks should be carefully removed and disposed of properly – see Chapter 7.

The main problem with this application is the view taken of the use of stored water for drinking purposes. The authors have considered the worst position, i.e. people will continue to drink stored water although it is strongly recommended they should not do so. With this point in mind it is essential that installations are completed by a close fitting but not airtight lid to reduce contamination of stored water.

There is little to choose between alternatives from an environmental perspective.

Alternatives	Technical Comment	Rank
1 Galvanized mild steel	Self supporting over bearers. Galvanized coating eventually broken down leading to corrosion.	2
2 Glass reinforced plastic (GRP)	Requires full boarded base for support. Plastic types easier to cut for pipework.	1
3 Poly-ethelene	Requires full boarded base for support. Plastic types easier to cut for pipework.	1
4 Poly-propylene	Requires full boarded base for support. Plastic types easier to cut for pipework.	1
5 Asbestos cement	No longer used in new work.	-

Health Comment	Rank	Environmental Issues	Rank	Cost Rank
No significant risk foreseen to occupants. (Unprotected gas cutting in loft space for disposal would present a fume fever risk.)	0/0	Steel could be recycled, but galvanised finishes may cause problems with this.	1/0/1/1	164
No significant risk foreseen to occupants.	0/0	Resin binders. Pollution during manufacture and disposal.	1/0/0/1	199
No significant risk foreseen to occupants.	0/0	Moderate embodied energy. Pollution from petrochemicals. Downstream score assumes recycling at end of life.	1/0/0/1	100
No significant risk foreseen to occupants.	0/0	Moderate embodied energy. Pollution from petrochemicals. Downstream score assumes recycling at end of life.	1/0/0/1	450
Aggressive water will release fibres. Although the effects of their ingestion is uncertain, on the grounds of prudence this continued use is to be avoided.	3/3	Asbestos is a hazardous waste and needs to be disposed of properly. Asbestos in existing buildings needs to be identified, recorded, assessed for comparative risks and removed or perhaps treated or encased and managed as appropriate. The (-) classification refers to lack of availability/no longer used.	-/-/-/3	N/A

Application 5.5 ELECTRICAL DUCTS AND CONDUITS	Alternatives	Technical Comment	Rank
Typical Situation cover–screw fixed clip fix cover cables socket plate cable solvent held joint (plastic) compression joint (metal) BOX DUCTING plastic systems PIPE CONDUCT metal and plastic systems	1 Galvanised steel	High strength with proven performance. Must be earthed.	1
	2 Un-plasticised Polyvinyl Chloride (PVCu)	Adequate performance. Tube and box systems easy to install. Risk of impact damage.	2

Technical Requirements

Protect and conceal electrical wiring from impact, abrasion and contact with occupants. Adequate strength to perform these functions. Ease of installation and decoration an advantage for exposed systems.

Decay and Degradation Factors

Impact. Water from plumbing and other leaks. Fair wear and tear. Decoration. Electrical systems require frequent extension and/or adaption causing damage to conduit.

Guidance Notes

Although galvanised steel offers high strength and impact resistance PVCu is adequate for domestic applications.

Neither options provide health hazards although asbestos as fire barriers or acoustic control may be found in larger trunking in multi-storey buildings, eg. between flats, which could put service workers unwittingly at risk.

Galvanised steel has a marginally lower environmental impact. In existing multi-storey buildings asbestos cement ducts may have been used and should be removed and disposed of in a managed way.

Health Comment	Rank	Environmental Issues	Rank	Cost Rank
No significant risk foreseen to occupants. (Extensive gas cutting and welding would present a fume fever risk.)	0/0	Steel could be recycled but galvanised finish causes problems. Good durability in intended applications.	2/0/0/1	160
No significant risk foreseen to occupants.	0/0	Problems from pollution during manufacture but situation is improving. Problems at disposal.	2/0/1/2	100

Application 5.6 FLUES AND FLUE PIPES	Alternatives	Technical Comment	Rank
Typical Situation liner · cement grout · insulation · flexible galvanised pipe flue wall · patent clip seal MASONRY FLUE with rigid liner · METAL insulated · STEEL LINER	1 Masonry/ clay liner	Traditional method of flue construction utilizing sectional linings requiring good workmanship for effective seal.	1
	2 Masonry with concrete liner	Traditional method of flue construction utilizing sectional linings requiring good workmanship for effective seal.	1
Technical Requirements Sealed non-combustible pipe to vent combustion gases to external air. Effective joint seals are absolutely essential. Span between recommended fixing centres. Adequate isolation from combustible structure. Resistance to chemical attack from combustion product and condensate to give adequate life. Must be durable and frost resistant where exposed externally.	3 Double wall stainless steel: mineral fibre filled	Generally convenient to fix. Good appearance.	2
	4 Double wall with galvanised steel outer wall and aluminium inner wall - air filled.	Generally convenient to fix. Good appearance. Normally used for gas and oil fired appliances. May not be suitable for solid fuel.	2
Decay and Degradation Factors External application – natural weathering. Frost. Wild life, impact damage. Wind. Internal application – possible decoration, normal wear and tear, impact. Interior of flue – attack by combustion and condensate.	5 Masonry with galvanised steel liner	Usually used for retrofit. Flexible. Joints only at top and bottom connections. Gauge/grade depends on class of fire appliance.	1
Guidance Notes Escape of flue gases from poorly constructed or badly maintained flues forms a much more serious hazard than that posed by the construction materials. The flue pipe options have to be carefully fitted to comply with fire requirements where they penetrate timber floors and roofs.	6 Asbestos cement	No longer used in new work. Existing flues need careful removal and disposal.	-

Guidance Notes

Escape of flue gases from poorly constructed or badly maintained flues forms a much more serious hazard than that posed by the construction materials. The flue pipe options have to be carefully fitted to comply with fire requirements where they penetrate timber floors and roofs.

Asbestos cement was extensively used in the past for this application. Normal replacement cycles for heating equipment means that most of this material will now have been replaced. Any remaining should be carefully removed, disposed of properly and the process managed - see Chapter 7.

From an environmental perspective metal flue pipes use less material overall but this is offset to some degree by lower life expectancy and the use of composite materials in their construction make them difficult to recycle.

Health Comment	Rank	Environmental Issues	Rank	Cost Rank
No significant risk foreseen to occupants.	0/0	Uses large volume of masonry material, with moderate embodied energy. Potential for recycling bricks and blocks – downstream '0' if recycled.	2/0/0/1	351
No significant risk foreseen to occupants.	0/0	Uses large volume of masonry material, with moderate embodied energy. Potential for recycling bricks and blocks – downstream '0' if recycled.	2/0/0/1	196
No significant risk foreseen to occupants. (The integrity of the casing requires to be preserved otherwise risk will be elevated by fibre release to 0/1.)	0/0	Score assumes UK source of stainless steel with 100% recycled content. Separation and disposal of M.M.M.F. insulation is problematic at the end of life.	1/0/1/1	345
No significant risk foreseen to occupants.	0/0	Composite construction hinders recycling. Impacts from galvanised coating on steel.	1/0/1/1	210
No significant risk foreseen to occupants.	0/0	Composite construction hinders recycling. Impacts from galvanised coating on steel, but galvanised steel linings can be less durable in some applications.	2/0/2/1	100
Fibre release on ageing, cleaning, maintenance and disposal presents a potential risk to maintenance workers, DIY occupants, construction and demolition operatives.	1/3	Asbestos is a hazardous waste and needs to be disposed of properly. Asbestos in existing buildings needs to be identified, recorded, assessed for comparative risks and removed or perhaps treated or encased and managed as appropriate. The (-) classification refers to lack of availability/no longer used.	-/-/-/3	N/A

Application 5.7
SANITARY FITTINGS

		Alternatives	Technical Comment	Rank
Typical Situation		1 Acrylonitrile, Butadiene Styrene (ABS)	Poorer scratch resistance. Lighter weight.	1
		2 Acrylic	Poorer scratch resistance. Lighter weight.	2
BATH WC/BIDET WASH HAND BASIN SHOWER TRAY		3 Enamelled cast iron and steel	Used for baths, heavy, well proven, durable.	2
		4 Pressed Steel	Well proven. Durable, medium weight.	1
Wash-hand basin Bath Sink Shower tray Or parts thereof		5 Vitreous China	Heavier, well proven. Durable.	1
		6 Fire Clay	Heavier, well proven. Durable.	1
Technical Requirements Watertight containers for a range of personal hygiene requirements. Ability to withstand loadings, impact and temperature changes. Ease of cleaning. Ease of installation.		7 Polymer Concrete (shower trays)	Heavier, durable. Can be purpose made for available space	1
Decay and Degradation Factors Chemical and abrasive cleaners. Temperature changes. Excessive loading and impact. Fair wear and tear.		8 Corian	Used for combined kitchen sinks and worktops.	1

Guidance Notes

Selection of alternatives is based mainly on appearance and colour requirements.

Ceramic and enamelled alternatives (4, 5 and 6) are heavier and provide more resistance to damage due to poor maintenance and cleaning. This is also reflected in the quantity of material and energy in manufacture. 'Plastics' based alternatives being lighter are generally easier to install.

Reused or reclaimed fittings supply the least environmental impact, provided the use of aggressive chemical cleaners is limited during renovation of the fittings.

Health Comment	Rank	Environmental Issues	Rank	Cost Rank
No significant risk foreseen to occupants.	0/0	Moderate embodied energy – pollution from petrochemicals. Downstream score assumes recycling.	1/0/1/0	Refer to specialist suppliers
No significant risk foreseen to occupants.	0/0	Moderate embodied energy – pollution from petrochemicals. Downstream score assumes recycling.	1/0/1/0	
No significant risk foreseen to occupants.	0/0	High embodied energy. Easily reused and/or refurbished enamel coating for reuse.	2/0/0/0	
No significant risk foreseen to occupants.	0/0	High embodied energy. Downstream score assumes recycling.	2/0/0/0	
No significant risk foreseen to occupants.	0/0	High embodied energy. Clay extraction. Easily reused, very durable.	2/0/0/0	
No significant risk foreseen to occupants.	0/0	High embodied energy. Clay extraction. Easily reused, very durable.	2/0/0/0	
No significant risk foreseen to occupants.	0/0	Moderate embodied energy. Unlikely to be reused.	1/0/0/1	
No significant risk foreseen to occupants.	0/0	Moderate embodied energy. Unlikely to be reused.	1/0/0/1	

Application 6.1 EXTERNAL PAVINGS	Alternatives	Technical Comment	Rank
Typical Situation 25 mm — tarmacadam on hardcore base well compacted 50 mm — slabs on cement dabs on compacted hardcore base bricks on rolled sand timber boarding on joists joists stakes driven into ground grass — hollow concrete "Grasscrete" blocks on sand base	1 Concrete (a) In situ (b) Slabs (flags) (including artificial stone) (c) Terrazzo	(a) Heavy duty. Requires formwork. Adaptable. Alternative surface finishes. Susceptible to stains. (b) Requires adequate bedding. (c) Requires adequate bedding. Visually used for patios or similar footways.	1 1 1
Technical Requirements Smooth paving to facilitate vehicle and pedestrian movement. Carry loadings. Resist frost. Rot resistance. Slip resistance. Chemical and oil resistance.	2 Brick (a) Clay (b) Concrete	(a) Heavy duty. Requires adequate side restraint and bedding. (b) Ditto	1 1
Decay and Degradation Factors Chemical and oil spills. Frost, movement of substructure. Normal weathering. Overloading. Abrasion from traffic. Mould growth, vegetation and wild life.	3 Tarmacadam	Poor appearance compared with others depending on aggregates. Requires adequate bedding. Attacked by oil and petrol spillage.	2
Guidance Notes A determining factor on the choice available is whether paving is purely for foot traffic or is also to carry vehicles. Road vehicles require thicker substrate, usually of hardcore (which is excluded in the environmental ranking. The impact of this can be reduced in certain locations by use of recycled aggregate – see Section 7.2). For this application cost is a good guide to durability and the quantity of maintenance required. In the U.K. timber is primarily used for decorative effect in footways and patios. Maintenance and DIY workers sawing or grinding materials in applications 1, 2b, 5 and 7, are at a potential health risk depending on the frequency with which they do the work, the amount of dust inhaled, and the crystalline silica content of the concrete aggregate, stone or cobble. Protective masks must be worn. From a general environmental perspective permeable solutions are to be preferred, eg. Options 4, 6 and 7, if boards are fixed with gaps. In terms of hard surface materials – reuse and recyclability is a key factor. Use of moderate strength mortars for spot bedding and jointing of units, or in sand as with option 2, facilitates recycling. Use of tropical hardwoods should be discontinued – see Chapter 3.	4 Loose stone (a) River gravel (b) Crushed rock 5 Slate and Stone slabs and cobbles 6 Timber boards on framing (a) Softwood (b) Temperate hardwood (c) Tropical hardwood 7 Grasscrete	(a) Needs good compaction subject to movement due to traffic. Becomes contaminated with dust and dirt over time. Natural drainage. (b) Ditto Heavy duty if properly bedded. Slightly more difficult to lay. Foot traffic only (a) Needs preservative treatment. (b) Some species durable in contact with the ground. (c) Ditto (footways only) Open concrete blocks which permits grass to grow through. Natural drainage. Can be heavy duty for special access areas.	4 4 1 1 1 1 2

Health Comment	Rank	Environmental Issues	Rank	Cost Rank
No significant risk foreseen to occupants.	0/0	(a) Impacts due to cement manufacture and aggregate extraction.	1/0/0/1	292
		(b) As above, but assuming bedding permits reuse, otherwise downstream = 1.	1/0/0/0	756*
		(c) Impacts due to cement manufacture and aggregate extraction.	1/0/0/1	1332++
No significant risk foreseen to occupants.	0/0	(a) High embodied energy and atmospheric emissions. Assumes sand bedding and reuse.	2/0/0/0	1060
		(b) Lower manufacturing energy. Assumes sand bedding and reuse.	1/0/0/0	828
No significant risk foreseen to occupants.	0/0	Pollution from petrochemicals. Durability very dependent upon workmanship.	1/0/1/1	684
No significant risk foreseen to occupants.	0/0	(a) Suspect long term durability due to gradual compression into the ground. Maintenance required.	0/0/1/0	269
		(b) Suspect long term durability due to gradual compression into the ground. Maintenance required.	0/0/1/0	100
No significant risk foreseen to occupants.	0/0	Reused New Both assume reuse at end of life.	0/0/0/0 1/0/0/0	N/A 3275+++
No significant risk foreseen to occupants.	0/0	(a) Available from sustainable sources. Environmental impacts from preservative treatment.	1/0/1/1	655
		(b) Availability from sustainable sources is limited. Upstream score is 1 if from sustainable sources.	2/0/0/0	1397
		(c) Large proportion of tropical timber is still sourced from virgin forest. Upstream score 1 if from sustainably managed sources.	3/0/0/0	1300
No significant risk foreseen to occupants.	0/0	Less material than other concrete options. Grass requires maintenance, eg. Usually by regular mowing.	0/0/1/0	990

* artificial stone
++ polished tiles
+++ stone (new)

Application 6.2 FLAT ROOF PROMENADE TILES	Alternatives	Technical Comment	Rank
Typical Situation	1 Glass reinforced cement (GRC)	Lightweight.	1
	2 Concrete slabs	Heavyweight compared with alternatives.	2
	3 Asbestos cement	Material commonly selected in the past due to combination of good durability and lightness in weight. No longer available but many still exist on older buildings.	-

Technical Requirements

Paving to cover walkways, etc, on flat roofs. Durable, rot and frost resistant material required. Lightweight types must permit bonding and be resistant to attack by bitumen solutions. Provide protection to the roof membrane from damage and solar radiation. Slip resistance important. Good fire properties (i.e. resistance to ignition) may be required.

Decay and Degradation Factors

Normal weathering. Frost. Fire. Mould growth, vegetation and wild life. Abrasion from maintenance and foot traffic.

Guidance Notes

Selection between types influenced by construction of flat roof deck and type of membrane utilized. For lightweight roof construction GRC has a major advantage. With heavyweight concrete decks selection is dictated to some degree by membrane design, e.g. heavyweight concrete slabs are required as ballast over insulation on inverted roof construction. The risk from existing sound asbestos cement tiles is not thought to be sufficiently serious to recommend immediate removal, which would be best left until major roof overhaul. Care should be taken on removal to preserve integrity of tiles. Dispose of properly in accordance with waste disposal regulations – see Chapter 7. Do not abrade for cleaning and maintenance.

There is little to choose from an environmental perspective, however GRC is lighter and therefore has lower structural requirements resulting in smaller sections in the structural decking.

Health Comment	Rank	Environmental Issues	Rank	Cost Rank
No significant risk foreseen to occupants.	0/0	Moderate embodied energy. Impacts due to cement manufacture.	1/0/1/1	Refer to specialist suppliers
No significant risk foreseen to occupants.	0/0	Impacts due to cement manufacture and aggregate extraction.	1/0/0/1	Refer to specialist suppliers
Fibre release on ageing, cleaning, maintenance and disposal presents a potential risk to maintenance workers, DIY occupants, construction and demolition operatives.	1/3	Asbestos is a hazardous waste and needs to be disposed of properly. Asbestos in existing buildings needs to be identified, recorded, assessed for comparative risks and removed or perhaps treated or encased and managed as appropriate. The (-) classification refers to lack of availability/no longer used.	-/-/-/3	N/A

Application 6.3
BOUNDARY FENCING AND WALLING

Typical Situation
Fencing for boundary and land sub-division. Live-stock control.

POST AND WIRE   POST AND RAIL   POST AND PANELS

Technical Requirements
Means of subdividing exterior spaces. Wind resistant, durable, maintenance free. Security and privacy may be required. Appearance important. Ease of installation may be important (especially DIY).

Decay and Degradation Factors
Frost, normal weathering, excessive wind loading, impact, vandalism, ground movement.

Guidance Notes
Technical ranking is shown within the main categories – eg. post and rail. It is assumed that the system is properly installed, eg. with proper concrete 'boots' for uprights and that softwood timber is treated with preservative, which has its own negative environmental impact. Selection is based on a balance between appearance, height, privacy, security and cost factors, together with other issues such as wind resistance requirements. Walls, particularly drystone walls, and hedgerows occupy more space.

From an environmental perspective use of hedging, as growing plants that contribute to the habitat for fauna and also fix CO_2, is to be preferred however one has to bear in mind some of the health implications as shown in the health comment. This option also requires regular maintenance which can be excessive with certain fast growing species, eg. Leylandi (and this can also poison neighbourly relations!).

Alternatives	Technical Comment	Rank
1 Post and Rail	Good wind resistance. Timber needs preservative treatment unless durable hardwood species selected. Timber easier to install and is adaptable in situ.	
a) Timber & timber		2
b) Concrete & timber		2
c) Concrete & concrete		1
d) PVCu & PVCu		4
2 Post & Panel	Poor wind resistance. Timber needs preservative treatment unless durable hardwood species selected. Accuracy in setting out and alignment of posts is necessary.	
a) Timber & timber		3
b) Concrete & concrete		1
c) Concrete & timber		2
3 Post & Wire	Good wind resistance but very poor privacy. High security systems available. Speed of erection varies with system selected but note need for tensioning of horizontal wires and the associated concrete foundations at the end of lines.	
a) Timber & galvanised steel wire		3
b) Timber & plastic coated gavanised steel wire		2
c) Concrete & galvanised steel wire		2
d. Concrete & plastic coated steel wire		1
e) Steel & galvanised steel wire		3
f) 'Steel' & plastic coated steel wire		2
4 Walling	Good durability, security and privacy depending on design and layout. Slow erection time. See also application sheet 7.3	
a) Bricks		1
b) Blocks		1
c) 'Dry' stone		1
d) 'Wet' stone		1
5 Hedges (Various)	Select species depending upon the key attributes, durability, maintenance security, privacy and appearance.	Rela-ted to spec-ies

Health Comment	Rank	Environmental Issues	Rank	Cost Rank
No significant risk foreseen to occupants.	0/0	Timber – impacts from preservative treatment. Concrete – mainly from cement manufacturer. PVC – problems from manufacture of polymer, but improving. Recycling of all above unlikely in practice.	1/0/1/1 1/0/0/1 1/0/0/1 2/0/1/1	280 308 336 317
No significant risk foreseen to occupants.	0/0	Timber – impacts from preservative treatment. Panels usually require high level of maintenance. Concrete – impacts mainly from cement manufacture. Recycling of all above unlikely in practice.	1/0/1/1 1/0/0/1 1/0/1/1	375 415 395
No significant risk foreseen to occupants.	0/0	Timber – impacts from preservative treatment. Concrete – impacts mainly from cement manufacture. Steel wire – contains metal smelting, galvanised coating and PVC coating. Recycling of all above unlikely in practice.	1/0/1/1 1/0/0/1 1/0/1/1 2/0/0/1 2/0/1/1 2/0/1/1	100 108 120 127 147 155
No significant risk foreseen to occupants.	0/0	Bricks – high embodied energy. Blocks – moderate embodied energy and cement manufacture. Dry stones – assume reused material – upstream = 1 if new quarried Wetstone – as drystone except little reuse anticipated at end of life.	2/0/0/1 1/0/1/1 0/0/0/0 0/0/0/1	801+ N/A 1102++ 1858+++
The leaves, bark or berries are toxic in certain plants used in hedging (eg. Laburnum, Cherry Laurel, Rhododendron, Yew). Privet is a popular hedging plant whose pollen induces asthma in some gardeners.	Vari-able	Greenest solution provided species selected does not require high level of energy intensive maintenance. Impacts in use – score would increase where powertools are regularly used.	0/0/0/0	Refer to specialist supplier

+ one brick thick: Commons; English bond; vesta
++ 200mm granite: Dry
+++ 450mm granite: Battering face

Application 7.1(a) CONCRETE AND ADDITIVES	Alternatives	Technical Comment	Rank
Typical Situation Applications in foundations. Ground Floor slabs. First Floor slabs. In Pre-cast such as floor beams, paving slabs, blocks and site works.	**Natural Aggregates +**		
	1 Ordinary Portland Cement (OPC)	Satisfactory for all applications except where option 3 is required. Special type produced for ready mix industry requiring less water.	
Technical Requirements Depends on application. Normally for in-situ work in dwellings, low crushing strength (20kn/mm^2. For pre-cast higher strengths will be necessary. Ground contact may require sulphate resisting cement depending upon soil conditions.	2 Rapid Hardening Cement (RHC)	Ditto above. More finely ground.	
Decay and Degradation Factors Frost. Water penetration. Inadequate cover to reinforcement at installation. Sulphates and other ground contaminants. Discolouration due to biological agents. Road salt may assist corrosion of rebar.	3 Sulphate Resisting Cement	Required where sulphates are present in the ground – seek advice from local building control. Darker in colour than 1 and 2.	
Guidance Notes Generally these comments refer to ready-mix, in-situ cast concrete. Technically mix design or selection must be related to purpose – for most applications in housing such as foundations and ground floor stability this means relatively low strengths (20kw/mm^2 casting strength). Therefore options are not ranked. From a health viewpoint, while unprotected cutting and drilling of concrete, whose aggregate has a high crystalline silica content, has produced lung disease, such a level and duration of exposure is unlikely to be experienced by maintenance and DIY workers.	4 No-Fines Concrete	No sand. Marginally superior insulation properties. Formally used in system housing for loadbearing external walls. Some production advantages.	
Skin contact with cement is a common cause of dermatitis, sometimes associated with sensitisation to chrome, to which maintenance and DIY workers could be at risk	**Man Made Aggregates +** 5 Ordinary Portland Cement (OPC)	Satisfactory for all applications except where option 3 is required.	
Environmental impacts from cement form a significant part of the overall impact of concrete, so in order to reduce this to a minimum attention should be given to the following factors:	6 Rapid Hardening Cement (RHC)	Ditto above. More finely ground.	
• Selection of appropriate mix design with the lowest cement content feasible to meet strength and durability requirements • Use of additives and/or PFA to partially replaced cement in the mix • Selection of cement from a modern low energy, low emission plant /continued over	7 Sulphate Resisting Cement	Required where sulphates are present in the ground – see advice from local building control. Darker in colour than 5 and 6.	

Health Comment	Rank	Environmental Issues	Rank	Cost Rank
				Refer to specialist suppliers
No significant risk foreseen to occupants.	0/0	High embodied energy. Inputs from aggregate extraction and cement manufacture. Upstream (3) if toxic chemicals used as fuel to fire cement. Downstream (2) if not recycled.	2/1/0/1	
No significant risk foreseen to occupants.	0/0	Ditto.	2/1/0/1	
No significant risk foreseen to occupants.	0/0	Ditto.	2/1/0/1	
No significant risk foreseen to occupants.	0/0	Ditto.	2/1/0/1	
No significant risk foreseen to occupants.	0/0	Slightly reduced impact from aggregate extraction.	1/1/0/1	
No significant risk foreseen to occupants.	0/0	Ditto	1/1/0/1	
No significant risk foreseen to occupants.	0/0	Ditto.	1/1/0/1	

Application 7.1(b) CONCRETE AND ADDITIVES	Alternatives	Technical Comment	Rank
Guidance Notes (continued) • Reduce waste by good materials management and procurement • Use man-made aggregates where suitable • Optimising the Ready-Mix Plant/aggregate source/site location relationship to minimise transport requirements Use of in-situ concrete also has impacts through the use of formwork and mould oils. Design and site practice should maximise use of formwork. Other considerations with Pre-cast concrete include the fact that current practice is to use high cement content and energy in steam/water-vapour curing in order to maximise mould use and early age strength. In some cases this can be seen as a trade-off in using less material in similar sections or in 'hollow' floor units creating less waste and in reduced erection time. The latter results in less impact from the construction phase on the environment around the site. Selection of aggregates is dealt with in Application 7.2.	**Man Made Aggregates +** 8 Pulverised Fuel Ash (PFA) with OPC/RHC	Mainly pre-cast in blocks and paving slabs, etc. May be autoclaved to improved properties and accelerate hardening. Can be aerated using aluminium powder or agitation to improve thermal insulation at lower crushing strength.	
	9 Ground Granulated Blast Furnace Slag (GGBS) with OPC/RHC	Ditto	
	10 Recycled Aggregates (crushed brick and/or concrete)	Needs to be free of gypsum, plaster and organic contamination such as earth and timber. Approved recycling processes are necessary to ensure adequate quality.	
	Additives 11 Air enhancing agents	Synthetic detergents and vinol resin derivatives. Increases frost resistance of concrete and mortar.	
	12 Accelerating Agents	Reduce setting time – rarely used.	
	13 Retarders	Lignosulphonic or hydroxylated-carboxylic derivatives. Used in ready-mixed concrete and mortar.	
	14 Plasticisers	Lignosulphonic or hydroxylated-carboxylic derivatives. Increase workability/strength without addition of further cement or to reduce cement content.	
	15 Super-plasticisers	Sulphonated melamine formaldehyde or naphthalene formaldehyde condensates. Further increase in workability fluidity over other plasticisers.	

Health Comment	Rank	Environmental Issues	Rank	Cost Rank	
				Refer to specialist suppliers	
No significant risk foreseen to occupants.	0/0	Reduced impacts as smaller quantities of cement utilised. Impacts from cement manufacture as 1.	1/1/0/1		
No significant risk foreseen to occupants.	0/0	Ditto.	1/1/0/1		
No significant risk foreseen to occupants.	0/0	Reduced impacts from extraction of aggregates. Impacts from cement manufacture as 1.	1/1/0/1		
No significant risk foreseen to occupants. (For maintenance workers and enthusiastic DIY-ers, the skin hazards of cements would be added to on unprotected contact with additives.)	0/0	Manufacturers are all using the same or similar materials for their products, but may blend them in different proportions. The active volume of each material within the set concrete is very small. 0:002% air enhancing agents to 0.03% for plasticisers. Since there is not a choice of materials for each use which in turn have different functions, it is inappropriate to draw environmental conclusions as normally found in this column. However, consideration might be given to the environmental implication of not using them. For example, using plasticisers may result in a slightly less embodied energy or material by reducing the cement content.	N/A		
Ditto.	0/0				
Ditto.	0/0				
Ditto.	0/0				
Ditto.					

Application 7.2 AGGREGATES	Alternatives	Technical Comment	Rank
Typical Situation Used as fill, hardcore bases, and for concrete.	**Naturally Occurring**		
	1 Marine sand and gravel	Washed to remove salt (calcium chloride)	
	2 Surface extracted sand and gravel	Some sands and gravel need to be washed/screened to remove loam. Commonly used material.	
Technical Requirements Clean, inert material intended to satisfy size and grading requirements of particular applications for fill, hardcore and concrete.	3 Crushed rock	Commonly used material.	
Decay and Degradation Factors Needs to be free of contaminants particularly organic materials which will decay causing voids.	4 Recycled a) Concrete b) Brick	Needs to be free of gypsum plaster and organic contamination such as earth and timber.	
Guidance Notes High volumes of material are customarily utilised, therefore transport and the building locality are critical factors.	5 Asphalt road/pavings	Reused in road resurfacing as a proportion of the mix.	
Selection of aggregates is therefore a matter of local availability and meeting use/performance requirements (eg. fill, bases, concrete).	**Man Made** 6 Pulverised Fuel Ash (PFA)	Mainly used in precast concrete and blocks (See Applications 7.1 and 7.3)	
From an environmental perspective the ranking assumes local source to minimise transport impacts, which in urban areas could be reclaimed crushed bricks, crushed concrete, etc, or other forms of clean industrial waste, eg. PFA. Downstream ranking is omitted as aggregates are either incorporated in concrete (see Application 7.1) or used as hardcore, which is difficult to reclaim at end of life and usually is abandoned in the ground. The exception to this is deep fill which is likely to remain uncontaminated with earth. It should be identified and reused where appropriate.	7 Steel Blast Furnace Slag	Ditto	

Health Comment	Rank	Environmental Issues	Rank	Cost Rank
				Refer to specialist suppliers
No significant risk foreseen to occupants.	0/0	Damage to marine environment.	2/0/-/-	
No significant risk foreseen to occupants.	0/0	Visual and environmental impacts of gravel pits.	2/0/-/-	
No significant risk foreseen to occupants.	0/0	Visual and environmental impacts of quarrying.	2/0/-/-	
No significant risk foreseen to occupants.	0/0	Selection reduces impact through reduction in demand for alternatives.	0/0/-/-	
No significant risk foreseen to occupants.	0/0	As above, but assumes large volumes on roads and paving.	0/0/-/-	
No significant risk foreseen to occupants.	0/0 0/0	Score assumes use in concrete and represents a balance between recycling an otherwise waste material against with energy required for processing.	1/0/-/-	
No significant risk foreseen to occupants.		Score assumes use in concrete and represents a balance between recycling an otherwise waste material against with energy required for processing.	1/0/-/-	

Application 7.3 BRICKS AND BLOCKS	Alternatives	Technical Comment	Rank
Typical Situation Small, manhandable units used dry or more usually with mortar to make walls or pavings.	1 Clay Bricks a. Ordinary solid b. Ordinary perforated c. Flettons d. Stocks (softmud)	Check crushing strength and frost resistance suitability for the application with manufacturer. Otherwise selection mainly determined by appearance factors.	
Technical Requirements Strength, dimensional stability and frost resistance appropriate to application. Sound and thermal properties may be important in walls. Good durability and low maintenance required. Mortar specification is a significant factor in determining performance in a number of applications, e.g. in external walls.	2 Calcium Silicate (sandline)	Check crushing strength and frost resistance suitability for the application with manufacturer. Otherwise selection mainly determined by appearance factors.	
Decay and Degradation Factors Frost, sulphates and other air and ground pollutants. Thermal and moisture movements. Excess moisture content.	3 Concrete bricks	Check crushing strength and frost resistance suitability for the application with manufacturer. Otherwise selection mainly determined by appearance factors.	
	4 Recycled bricks	Most bricks are capable of recycling. Strong cement mortars militate against recycling at end of life.	
Guidance Notes Key selection factors are the appearance and durability. The latter is vitally important in loadbearing buildings as incorrect selection may result in extensive repair work with additional environmental impacts during the building life and in extreme cases in premature demolition. Mortar strength should be weaker than the brick or block units to ensure good durability and to aid recycling at end of life. Reduced cement content or lime mortars also help reduce the environmental impacts of cement manufacture. Attention should be given to good practice in overlapping eaves, cills and in damp-proofing walls to protect from frost and water erosion. Options 7, 8 and 9, and in some locations 1(c), are used internally or are rendered or otherwise protected (eg. with tile hanging) when used externally.	5 Natural Stone	Durability related to source quarry, method of extraction, and exposure factors – seek advice from supplier.	
	6 Recycled natural stone	Contain very dense and durable stones, eg. Granite and gritstone are eminently recyclable, others such as limestone or sandstone may not prove durable when recycled or may need re-dressing.	
From an environmental perspective optimising the supplier/site location relationship to minimise transport requirements is a key factor. The upstream ranking assumes 'local' supply. Lighter weight units offer an advantage, and in some blocks offer improved thermal performance. However the contribution of dense bricks and blocks to the thermal mass of the building, assisting in passive cooling and solar design, should not be overlooked.	7 Aggregate Concrete Blocks a Steam cured b Autoclaved c Air cured d Artificial Stone	Check crushing strength and frost resistance suitability for the application with manufacturer. See application sheet 7.1 and 7.2.	
Recycled units offer the lowest impact but require close visual inspection to ensure reasonable durability. Otherwise, natural stone and concrete provide the minimal environmental impact. The downstream ranking assumes that only a small percentage are reused, but this could approach zero if a high percentage of re-use became more common practice.	8 Aerated Concrete	Lightweight compared to other materials. Improves thermal performance. 7kn/mm2 minimum crushing strength below dpc.	
	9 Recycled blocks	Blocks are difficult to recycle – there is a tendency for the block to break rather than the mortar joint on demolition.	

Health Comment	Rank	Environmental Issues	Rank	Cost Rank
No significant risk to occupants foreseen.	0/0	a. High embodied energy and emissions. Impacts from clay extraction.	2/0/0/1	200*
		b. High embodied energy and emissions. Impacts from clay extraction.	2/0/0/1	400*
		c. High embodied energy and emissions. Impacts from clay extraction.	2/0/0/1	540*
		d. Lower embodied energy is offset by higher emissions.	2/0/0/1	700*
No significant risk to occupants foreseen.	0/0	High embodied energy and sand/silicate extraction.	2/0/0/1	N/A
No significant risk to occupants foreseen.	0/0	Moderate embodied energy pollution from cement manufacture.	1/0/0/1	100*
No significant risk to occupants foreseen.	0/0	May not provide adequate durability in exposed positions. There is a limit to the number of re-uses (two?).	0/0/0/1	N/A
No significant risk to occupants foreseen.	0/0	Visual and environmental impact of quarrying.	1/0/0/1	3250*
No significant risk to occupants foreseen.	0/0	May not provide adequate durability in exposed positions. There is a limit to number of re-uses.	0/0/0/1	2400*
No significant risk to occupants foreseen.	0/0	Various embodied energy profiles. Pollution from cement manufacture. Some aggregates, eg. Unsintered PFA and furnace clinker, lower embodied energy and are recycled from other processes – downstream score 0. Upstream score 2 for artificial curing.	1/0/0/1	510*
No significant risk to occupants foreseen.	0/0	Higher embodied energy than 7. Aluminium used as foaming agent.	2/0/0/1	740*
No significant risk to occupants foreseen.	0/0	Problematic to re-use depending on mortar strength.	0/0/0/1	N/A

* supply only

Application 7.4 ADHESIVES	Alternatives	Technical Comment	Rank
Typical Situation Structural bonding a wide range of materials	1 Acryclic glues (acrylics and acrylates)	Used for wall tile and flooring tiles and in some fillers and sealants. Easier to use than traditional cementious mixes.	
Technical Requirements To attach or bond one material to another to withstand the working loads and give adequate workability time. Resistance to movement or ability to withstand movement and, in certain cases, moisture.	2 Cyano-acrylates	Known as super glues. Used for bonding two surfaces in tight contact. Glues most materials together; low impact resistance.	
Decay and Degradation Factors Moisture, frost (if external application), fungi. Normal wear and tear. Moisture and thermal movement.	3 Epoxy resins	2-part system – resin and curing agent. Used in grouts and fillers for strong adhesion. Can be used for floor tiles.	
Guidance Notes Adhesives are generally formulated for a particular purpose and therefore are not directly comparable from a technical standpoint. Speed of set may be an important buildability issue to either give adequate working time for minor adjustment, or to provide a rapid fix. From a health standpoint, option 5 is preferred to option 4 for internal applications, such as in man-made boards. The synthetic polymeric adhesives will react with healthy skin and after repeated use will lead to disabling dermetitis. Safe working techniques are necessary to protect DIY and construction workers from contact. Most of the alternatives have a similar environmental impact. Therefore the main consideration is to use a minimum of material of adequate bonding strength sufficient to do the work. This will be significant at end of life to ease separation and dismantling for recycling. For this reason, for a range of circumstances, it may be better to select mechanical fixings in preference.	4 Formalde-hyde resins	Binder in chipboard. Strand board, MDF and plywood.	
	5 Isocyanate resins	Increasingly used as non off-gassing gluing substitute for formaldehyde resins.	
	6 Polyvinyl acetate PVA	Used for on-site jointing. Also in some wall tile adhesives and heavy duty wallpaper glues.	
	7 Traditional and natural glues	Internal use only. Usually for joinery work.	

Health Comment	Rank	Environmental Issues	Rank	Cost Rank
No significant risk foreseen for occupants.	0/0	High energy in production. Using materials from non-renewable resources. Pollutants from petrochemical manufacture. Hinders recycling process of other materials and therefore adds to quantity of waste.	2/0/0/2	Refer to specialist supplier
No significant risk foreseen for occupants.	0/0	High energy in production. Using materials from non-renewable resources. Pollutants from petrochemical manufacture. Hinders recycling process of other materials and therefore adds to quantity of waste. Only used in relatively small amounts.	2/0/0/0	
No significant risk foreseen for occupants.	0/0	High energy in production. Using materials from non-renewable resources. Pollutants from petrochemical manufacture. Hinders recycling process of other materials and therefore adds to quantity of waste. Only used in relatively small amounts.	2/0/0/2	
Volatile unreacted re-agents, if they build up in the indoor environment, will lead to irritation of eye, nose and throat. Formaldehyde is a suspect carcinogen . Release indoors is unacceptable. (3/3)	0/0	High energy in production. Using materials from non-renewable resources. Pollutants from petrochemical manufacture. Use of material contributes to problem of disposal of man made boards. Incineration may release harmful gases. Urea formaldehyde tends to contain most free formaldehyde.	2/0/0/2	
Volatile unreacted re-agents, if they build up in the indoor environment, will lead to irritation of eye, nose and throat.	0/0	High energy in production. Using materials from non-renewable resources. Pollutants from petrochemical manufacture. Use of material contributes to problem of disposal of man made boards.	2/0/0/2	
No significant risk foreseen for occupants.	0/0	High energy in production. Using materials from non-renewable resources. Pollutants from petrochemical manufacture. Subject to microbial degradation.	2/0/0/1	
No significant risk foreseen for occupants.	0/0	High energy to manufacture but usually natural products. If made with synthetic components upstream 2.	1/0/0/0	↓

Application 7.5 SEALANTS	Alternatives	Technical Comment	Rank
Typical Situation Sealing joints. eg. expansion joints, around windows and doors, floors and walls, structural panels and sanitary goods.	1 Preformed strips	Simplest type of sealing compound. Only suitable for smooth, non porous surfaces. eg, glass. Those containing oil can cause staining in porous materials. Movement accommodation is poor. Ranking depends upon composition, see options 2, 3 and 4 below.	
Technical Requirements Providing a watertight and/or airtight seal at various junctions and joints. Accommodate movement, ability to bridge the gap, resist chemical attack. Appearance important for external use. Should accept decoration. Weather resistant. Maintain a water and/airtight seal.	2 Mastics:		
	a) Oil based (oleo resinous)	Limited movement accommodation. Typical life expectancy 5 years.	4
Decay and Degradation Factors UV light, wildlife (especially birds). Natural weathering, chemical attack. Frost. Normal wear and tear.	b) bitumen based	Very good waterproofing properties, can be used below ground. Has good resistance to microbial and chemical attack. Life expectancy 10 years.	3
Guidance Notes The technical ranking considered external applications and is mainly based on durability factors. Preformed strips are easier to fit in some circumstances, eg. to seal joints between lightweight cladding sheets.	c) butyl rubber based	More flexibility than oil based. UV degradation a problem.	3
As a generality, the quantity of these materials and their sites of installation around baths, doors and window frames, present no significant hazard in-situ to adult occupants. Infants and young children with pica swallow non-nutritious materials found around the house including bits of its fabric. As a consequence, there are standards covering the wholesomeness of toys, domestic paint, putty and plumbing sealants.	3 Semi elastomeric sealants	Superior to mastics, especially movement accommodation and durability. Applied from cartridges. Perform well in slow moving joints. Can attract dust. Life expectancy 10-15 years.	
Durability is also important in terms of environmental impact, as some materials may require replacement up to six times during the equivalent life of the best performing alternative. Otherwise in terms of initial application the volume of material in any individual dwelling is small.	a) Acrylics-emulsion based	Commonly used for internal sealing of window and door frames, gaps between ceilings and walls, etc. Can be used externally.	2
	b) Acrylics-solvent based	Primarily used for sealing externally.	2
	4 Elasto-meric sealants (silicone based)	The best performance. Highest movement accommodation. Weathering and durability. Used for heavy duty situation and expansion joins. Life expectancy 20-30 years.	1

Health Comment	Rank	Environmental Issues	Rank	Cost Rank
No significant risk foreseen for occupants.	0/0	Petrochemical manufacture and pollutants. Difficult to recycle. Problem at disposal.	1/0/1/1	Not comparable for cost ranking
No significant risk foreseen for occupants.	0/0	Main environmental impacts from oil, bitumen and butyl rubber. Manufacturing pollution and high embodied energy.	1/0/2/1	140*
			1/0/1/1	100*
			1/0/1/1	N/A
No significant risk foreseen for occupants.	0/0	Pollutants from petrochemical manufacture. High embodied energy. Potential hazard from disposal of unused materials.		
			1/0/1/1	429*
			1/0/1/1	470*
No significant risk foreseen for occupants.	0/0	High embodied energy. Pollutants from manufacture. Good durability.	1/0/0/1	409*

* supply only

Application 7.6 TIMBER	Alternatives	Technical Comment	Rank
Typical Situation Structural and constructional applications. Exterior joinery: eg, external doors, weatherboarding. Interior joinery: eg, internal doors, skirting boards, architraves.	**Structure and construction:** 1 Softwoods, eg. Douglas Fir, Scots Pine, Sitka Spruce, Southern Pine	Generally susceptible to decay if exposed to damp conditions without some form of protection or preservation.	
	2 Temporate hardwoods, eg. Oak, Elm	Durable, resistant to decay.	
Technical Requirements Structure and construction:- durability, strength, absence of structural defects, dimensional stability. Exterior joinery:- durability and resistance to decay, appearance and/or ability to receive decoration, dimensional stability. Interior joinery:- appearance and/or ability to receive decoration, dimensional stability.	3 Tropical hardwoods, eg. Keruing, Taun, Vitex	Durable, resistant to decay.	
Decay and Degradation Factors Dampness. Fungal and insect attach. Fire. Physical impact and abrasion of decorative timber. Wear and tear.	**Exterior joinery*:** 4 Softwoods, eg. Douglas Fir, Western Red Cedar, Redwood, Lorch	With exception of cedar, needs regular/frequent maintenance and/or preservative treatment to maintain durability.	
Guidance Notes Technically the main selection factor is whether timber is structural (ie. floor joist) or decorative (ie. skirting board). For most decorative timber applications, appearance will be the primary factor (eg. varnished floor-boards). In all cases the technical comments assume good building practice to protect the timber from significant dampness. Some species of hardwoods are durable in damp conditions, eg. Greenheart, but this parameter has not been specifically considered in this data sheet.	5 Temperate hardwoods, eg. White Oak (US), Oak (English), Sweet Chestnut	Durable.	
From a health perspective the main concerns arise from the dust generated from machinery. In the event of refurbishment or DIY activities involving sanding that generates fine dust, its inhalation may give rise to irritation of eyes, ears, throat and chest. This occurrence depends on the variety of wood, size and dose of particles, and susceptibility. Wood dust is also a suspect carcinogen. While the hardwoods generally have a reputation for producing these effects, softwoods are not exempt, Western Red Cedar being a well know example.	6 Tropical hardwoods, eg. Teak, Meranti, Vitex, Afzelia, Agba, Iroko, Mahogany	Durable.	
Some surface treatment products may cause dermatitis and breathing complaints.	**Interior joinery*:** 7 Softwood, eg. Variety of pines, Hemlock, Cedar, Spruces, Yew	Selection on workability factors and appearance for detailed applications.	
Timber should be selected from sustainably managed forests, but at present it remains very difficult to determine the exact source of the vast majority of timber available in the U.K. Chapter 3 provides details of the international situation on certification of timber. Due to the current rate of depletion of the tropical forest and the fact that the majority of tropical timber does not come from sustainable managed sources, great caution should be exercised when considering the use of tropical timbers in order to ensure sustainable source, See chapter 3.	8 Temperate hardwoods, eg. White Oak (US), Oak (European), Sweet Chestnut, Ash	Selection on workability factors and appearance for detailed applications.	
The main environmental impacts occur upstream, from transport and kiln drying. It is better to use local sources and the ranking assumes this to be the case. Whilst natural seasoning is to be preferred it is accepted that it would be difficult to meet current demand for softwoods using natural drying.	9 Tropical hard-woods, eg. African Walnut, Iroko, Mahogany, Meranti, Teak, Sapele, Taun	Selection on workability factors and appearance for detailed applications.	
Timber could be reused/recycled many times, but current constructional detailing and practice restrict this, which requires more attention from all sectors of the industry.	10 Recycled	Slight risk of attack from insects and fungi. Good standard of inspection required to identify species and defects. Otherwise as above.	

Health Comment	Rank	Environmental Issues	Rank	Cost Rank
No significant risk foreseen to occupants.	0/0	1. Upstream score 1 if source is not sustainably managed or transport to UK over large distance (eg. From outside Europe). Use phase score 1 if significant risk of attack by fungi or wood destroying insects. Downstream score 1 if burned or if timber contained preservative disposed to landfill.	0/0/0/0	100*+
No significant risk foreseen to occupants.	0/0	2. Upstream score 1 if source is not sustainably managed or transport to UK over large distance (eg. From outside Europe). Use phase score 1 if significant risk of attack by fungi or wood destroying insects. Downstream score 1 if burned.	0/0/0/0	530*◆
No significant risk foreseen to occupants.	0/0	3. Upstream score 1 if from sustainable managed sources. Downstream score 1 if burned.	3/0/0/0	326*◆◆
No significant risk foreseen to occupants.	0/0	4. Upstream score 1 if source is not sustainably managed or transport to UK over large distance (eg. From outside Europe). Downstream score 0 if timber reused/recycled.	0/0/1/1	184*++
No significant risk foreseen to occupants.	0/0	5. Upstream score 1 if source is not sustainably managed or transport to UK over large distance (eg. From outside Europe). Downstream score 0 if timber reused/recycled.	0/0/0/1	530◆*
No significant risk foreseen to occupants.	0/0	6. Upstream score 1 if from sustainable managed sources. Downstream 0 if timber reused/recycled.	3/0/0/1	326◆◆*
No significant risk foreseen to occupants.	0/0	7. Upstream score 1 if source is not sustainably managed or transport to UK over large distance (eg. From outside Europe). Downstream score 0 if timber reused/recycled.	0/0/0/1	184++*
No significant risk foreseen to occupants.	0/0	8. Upstream score 1 if source is not sustainably managed or transport to UK over large distance (eg. From outside Europe). Downstream score 0 if timber reused/recycled.	0/0/0/1	530*◆
No significant risk foreseen to occupants.	0/0	9. Upstream score 1 if from sustainable managed sources. Downstream 0 if timber reused/recycled.	3/0/0/1	326*◆◆
		Downstream score 1 if burned or if timber containing preservative disposed to land fill.	0/0/0/0	N/A

* supply only
+ carcassing software
++ joinery softwood
◆ American white oak
◆◆ Iroco

Application 7.7 TIMBER PRESERVATIVES	Alternatives	Technical Comment	Rank
Typical Situation Preservation of timber in structural members, external windows, joinery and other exposed situations.	1. Creosote	External use only, mainly for fencing and outbuildings. Bleeds over time.	
Technical Requirements Preservation by fungicidal and/or insecticidal action. Longterm effectiveness. Remain in timber, resisting washing and leaching out for as long as possible. Must not increase combustibility of timber. Coloured dyes may be required to identify preserved timber or to supply decorative effect.	**Water-based:** 2. Boron 3. Copper, Chrome, Arsenic (CCA)	Comments apply to options 2-10 below: Various formulations available from manufacturers depending upon degree of insecticidal or fungicidal action required and whether the material is to be brushed, sprayed, dipped or vacuum-impregnated.	
Decay and Degradation Factors Fire. Evaporation of volatile chemicals. Abrasion from sanding down where decoration is expected. Frequent human contact inevitable with decorative wood stains, but restricted with structural timbers. Wear and tear in exposed locations.	**Solvent-based:** 4. Zinc Napthenate 5. Copper Napthenate	Vacuum-impregnation is generally considered to give superior penetration.	
Guidance Notes It is important to remind designers and constructors of the need for good constructional practice to reduce the risk of dampness and therefore of fungal attack. The use of preservatives in this circumstance should be seen as an extra precaution and not as the primary means of prevention of decay. It should not be forgotten that these solutions are poisonous and therefore their use needs to be carefully evaluated. Variation in formulation between manufacturers products makes comment upon types difficult but from a health viewpoint the degree to which the preservative is fixed of locked in the timber is important. Occupants must be decanted while treatment is in progress and considerably increased ventilation for an appropriate period thereafter (eg. 2 weeks). In new buildings factory-treated timber would only cause concern when large quantities of timber are to be exposed within habitable rooms, e.g. if walls, ceilings, and perhaps floors are all faced with treated timber and this is combined with low ventilation rates. Designers should therefore be conscious of the quantity of treated timber incorporated into new construction and the possible effect this may have on the indoor air quality. While it may be necessary to treat structural timber in known risk areas to ensure durability, especially in timber framed construction, the need to treat decorative timber with volatile preservatives should be questioned. From an environmental perspective preservative treated timber presents a problem for disposal and this is reflected in the ranking. From this perspective options 2, 4 and 5 cause less concern. The score for the construction phase assumes that timber if pre-treated off-site. If not, increase score to 1. In all cases treated timber should not be burned as a means of disposal.	6. Pentochlor-Phenol (PCP) 7. Tributyl Tin Oxide 8. Lindane (Gamma HCH) 9. Permethrim 10. Acypetacs Zinc	Long-term effectiveness expected to be 30 years minimum (essential to confirm with manufacturers).	

Health Comment	Rank	Environmental Issues	Rank	Cost Rank
1. No significant risk foreseen to occupants. However, it is a complex mixture of some 160 biologically active chemical compounds, which can irritate and sensitise the skin to light, and cause cancer. The hazard is for maintenance and DIY workers with heavy skin contact.	0/0	1. Pollution from coal tar processes. Injuries to some forms of plant life.	2/0/2/2	Refer to specialist supplier
2. No significant risk foreseen to occupants.	0/2	2. Generally recognised as least toxic option of current preservation treatments available.	1/0/1/1	
3. Hexavalent chromium compounds and arsenic and its compounds are carcinogenic, and so safe levels of exposure have been recommended. As they are non-volatile, occupants are not at risk; only DIY workers are at risk if they apply the product or if they sand impregnated timber.	0/2	3. Arsenic is classed as a deadly poison. Serious problems at disposal of treated timber.	2/0/1/3	
4. No significant risk foreseen to occupants if correctly applied during construction. Subsequent use by maintenance or DIY workers may expose them and occupants to an elevated risk because of hazards from volatiles.	0/2	4. Solvents. Lower risk at disposal.	2/0/1/1	
5. No significant risk foreseen to occupants if correctly applied during construction. Subsequent use by maintenance or DIY workers may expose them and occupants to hazards from volatiles.	0/2	5. Solvents. Lower risk at disposal.	2/0/1/1	
6. It is reasonably foreseeable that the level of containment required to ensure that occupants of a newly treated building are free of risk might not regularly be attained. Likewise, its use subsequently by maintenance or DIY workers is likely to expose them and occupants to an elevated risk.	1/3	6. Solvents. Incineration at disposal results in highly toxic fumes.	3/0/1/3	
7. No significant risk foreseen to occupants. Subsequent use by maintenance or DIY workers may expose them to toxic hazard.		7. Solvents. Now banned as a boat anti-fouling paint.		
8. It is reasonably foreseeable that the level of containment required to ensure that occupants of a newly treated building are free of risk might not regularly be attained. Likewise, its use subsequently by maintenance or DIY workers may expose them and occupants to an elevated risk.	0/2	8. Solvents. Main cause of decline in bat colonies. Incineration at disposal results in highly toxic fumes.	3/0/1/3	
9. It is reasonably foreseeable that the level of containment required to ensure that occupants of a newly treated building are free of risk might not regularly be attained. Likewise, its use subsequently by maintenance or DIY workers is likely to expose them and occupants to an elevated risk.	1/3	9. Solvents. Safer for bats than 8.	3/0/1/3	
10. A relatively new compound for which lower toxity levels are claimed. At the time of writing it has not proved possible to confirm or deny this claim, therefore it has not been ranked. Use by maintenance or DIY workers may expose them and occupants to hazards from volatiles.	1/3	10. Solvents. Relatively new compound. Lower risk at disposal?	2/0/1/2	
			2/0/1/1	

Application 7.8 WATER BASED WALL AND CEILING PAINTS	Alternatives	Technical Comment	Rank
Typical Situation Decorative paint film for walls and ceilings are often refered to at emulsion paints.	**Interior First Decoration Products:** 1 Matt (Maximum VOC level: < 30g/l)	Sheen level <5% @ 60° Designed to be highly porous to allow fresh plaster substrates to dry thoroughly without disrupting the paint film.	
Technical Requirements Protective film applied to building materials to provide a decorative effect. Applied as a liquid (may be structured). Must have a good bond to the substrate. Good durability required in normal service.	**Redecoration Products:** 2 Matt (Maximum VOC level: < 30g/l)	Sheen level <5% @ 60°. Flat paint designed to disguise imperfections in the substrate.	
Decay and Degradation Factors Interior – Normal wear and tear from contact with occupants, condensation, etc. Exterior – Natural weathering processes. UV, water, fungal and algal attack, frost, etc.	3 Soft Sheen (Maximum VOC level: < 30g/l)	Sheen level <25% @ 60°. Mid sheen paint.	
Guidance Notes Selection of appropriate paint systems is dependent on the conditions of service of the film, the desired finish and the state of the substrate. Specifiers are encouraged to use a manufacturer's paint system to ensure multicoat compatibility. Generally two coats of the finish are recommended to minimise problems of uneven application.	4 Silk (Maximum VOC level: < 150g/l)	Sheen level >25% @ 60°. Sheen paint designed to highlight substrate relief. Particularly suitable for use with relief wallpapers.	
Health and Safety: All water based paints are effectively lead free – that is, they contain only background levels of lead, no lead is added during manufacture. The solvent levels, volatile organic compounds (VOC), quoted are the industry agreed maximum levels for the end of 1998. A further staged reduction has been agreed but the target date for compliance is still under discussion. There are many products available currently which are well below the quoted maxima, eg. Crown Solo Emulsion (soft sheen VOC <3g/l).	5 Acrylic Eggshell (Maximum VOC level: < 150g/l)	Sheen level >25% @ 60°. Designed for use in high wear areas, and areas subject to condensation, etc, eg. kitchens, bathrooms, hospitals, schools.	
Contamination of skin and clothing with fungicide containing paint, will constitute a hazard to the redecorator. Construction score 2 if excessive waste paint disposed of incorrectly. Paints have a relatively high embodied energy in relation to mass and that they have a relatively short life compared to other construction elements.	**Exterior:** 6 Textured Masonry Paints (Maximum VOC level: < 60g/l)	Restricted spreading rate designed to lay down a thick protective film. More prone to dirt pick up than smooth paints. Often carry BBA Certificates for 10-15 yr durability. Usually contains a film fungicide to protect against microbiological attack.	
	7 Smooth Masonry Paints (Maximum VOC level: < 60g/l)	Enhanced spreading rate. Usually highly flexible. Often carry a BBA certificate for 10 yr durability. Usually contains a film fungicide to protect against microbiological attack.	

Health Comment	Rank	Environmental Issues	Rank	Cost Rank
No significant risk foreseen to occupants.	0/0	Pollution from manufacture. High embodied energy.	1/0/1/0	105*
No significant risk foreseen to occupants.	0/0	Pollution from manufacture. High embodied energy.	1/0/1/0	100*
No significant risk foreseen to occupants.	0/0	Pollution from manufacture. High embodied energy.	1/0/1/0	147*
No significant risk foreseen to occupants.	0/0	Pollution from manufacture. High embodied energy. High solvent (Volatile Organic Compound) levels.	1/1/1/0	105*
No significant risk foreseen to occupants.	0/0	Pollution from manufacture. High embodied energy. High solvent (Volatile Organic Compound) levels.	1/1/1/0	176*
No significant risk foreseen to occupants.	0/0	Pollution from manufacture. High embodied energy.	1/0/1/0	100*
No significant risk foreseen to occupants.	0/0	Pollution from manufacture. High embodied energy.	1/0/1/0	117*

* supply only

Application 7.9 TRIM PAINTS	Alternatives	Technical Comment	Rank
Typical Situation Decorative paint film on doors, skirting boards, architraves, windows, etc., collectively known as trim.	**Water based Primer:** 1 Primer or Primer/Under-coat (Maximum VOC Levels: < 150g/l)	Designed to provide a strong bond to the substrate and be flexible enough to follow dimensional changes in the substrate caused through environmental stress, eg. temperature and humidity changes. Many comply with the BS5082 standard.	
Technical Requirements Protective film applied to trim to provide a decorative effect. Applied as a liquid (may be structured). Must have a good bond to the substrate. Good durability required in normal service. Also to protect external timber from dampness and decay.	**Finish:** 2 Satin (Maximum VOC Levels: < 150g/l)	Sheen > 25% @ 60°. Subdued gloss designed to disguise substrate defects. Highly flexible interior product.	
Decay and Degradation Factors Interior – Normal wear and tear from contact with occupants, condensation, etc. Exterior – Natural weathering processes. UV, water, fungal and algal attack, frost, etc.	3 Gloss (Maximum VOC Levels: < 150g/l)	Gloss typically > 50% @ 20°. Highly flexible gloss finish. Exterior versions usually contain a film fungicide to protect the paint from microbiological attack.	
Guidance Notes Selection of appropriate paint systems is dependent on the conditions to service of the film, the desired finish and the state of the substrate. Specifiers are encouraged to use a manufacturer's paint system to ensure multi-coat compatibility. Health and Safety: All water and solvent based decorative paints are effectively lead free – that is, they contain only background levels of lead, no lead is added during manufacture. The solvent levels quoted are the industry agreed maximum levels for the end of 1998. A further staged reduction has been agreed but the target date for compliance is still under discussion. Prolonged exposure of decorators to paint solvents, volatile organic compounds (VOC), has been linked with intellectual deterioration: this has given an added incentive to reduce toxin levels in paint. Occupation of freshly painted spaces inadequately ventilated will cause eye, nose and throat irritation. Therefore it is better to use waterbased paints. Durability assessment depends upon correct application in appropriate weather conditions. Environmental ranking of construction phase score 2 if excessive waste paint is disposed of incorrectly. (eg. into drains). Paints have a relatively high embodied energy in relation to mass and that they have a relatively short life compared to other construction elements.	**Solvent based** 4 Primer or Primer/Under-coat (Maximum VOC Levels: < 750g/l)	Designed to provide a strong bond to the substrate and be flexible enough to follow dimensional changes in the substrate caused through environmental stress, eg. temperature and humidity changes.	
	5 Gloss (Maximum VOC Levels: < 400g/l)	Gloss typically > 50% @ 20°. Highly flexible gloss finish.	

Health Comment	Rank	Environmental Issues	Rank	Cost Rank
No significant risk foreseen to occupants.	0/0	Pollution from manufacture – high embodied energy. Life expectancy score 2 for external applications.	1/1/1/0	100*
No significant risk foreseen to occupants.	0/0	Pollution from manufacture – high embodied energy. Life expectancy score 2 for external applications.	1/1/10	119*
No significant risk foreseen to occupants.	0/0	Pollution from manufacture – high embodied energy. Life expectancy score 2 for external applications.	1/1/1/0	127*
No significant risk foreseen to occupants when all fumes are ventilated.	0/0	Pollution from manufacture – high embodied energy. Life expectancy score 2 for external applications. Higher levels of solvent (Volatile Organic Compounds).	1/2/1/0	124*
No significant risk foreseen to occupants when all fumes are ventilated.	0/0	Pollution from manufacture – high embodied energy. Life expectancy score 2 for external applications. Higher levels of solvents (Volatile Organic Compounds).	1/2/1/0	140*

* supply only

Application 7.10 VARNISH, WOOD DYES, WOOD STAINS, DECORATIVE EFFECTS	Alternatives	Technical Comment	Rank
Typical Situation Varnishes, wood dyes and decorative effect products for all interior woodwork such as trim, panelling and floors. Woodstains for all exterior joinery, cladding and quality garden timber. Decorative fence and shed treatments for general purpose garden timber.	**Varnishes**	Generally clear coatings which are virtually impermeable, ie. stop liquid water entering the wood.	
	1. Water based (Max VOC levels: <150g/l)	Good performance on vertical surfaces.	
	2. Solvent based (Max VOC levels: <550g/l)	Good performance, tolerates poor substrate and poor joinery detailing.	
Technical Requirements Protective film applied to wood to provide a decorative semi-transparent (transparent in the case of unpigmented varnishes), which enhances the natural grain of wood. Applied as low/medium viscosity liquids, but can be structured. Must have good bond to the substrate. Good durability required. Interior products should have good resistance to common household chemicals and flooring finishes. Should have good hardness and abrasion resistance. Exterior products require good flexibility and (film) resistance to fungal/algal attack.	**Woodstains**	Generally coloured and opaque to a greater or lesser extent. Much more permeable than varnishes.	
	3. Water based (max VOC levels: <170g/l)	Good performance on vertical surfaces.	
	4. Solvent based (max VOC levels: <500g/l)	Good performance, tolerates poor substrate and poor joinery detailing.	
Decay and Degradation Factors Interior: normal wear and tear from usage, condensation and household cleaning (chemicals), etc. Exterior: natural weathering processes, UV, water, fungal and algal attack, frost, etc.	**One pack coatings for specific end use application**	For example: Floorboard coatings/seals.	
Guidance Notes Selection of appropriate woodcare system is dependent on the conditions of service of the film, the desired finish and the state of the substrate. If any form of pretreatment is required, eg. wood dye to change the natural colour of the substrate, specifiers/users are encouraged to use the same manufacturer's products to ensure compatibility. A three coat application is generally recommended, particularly on floors. Health and Safety: The VOC (volatile organic compound) levels are those agreed by the members of the British Coatings Federation Decorative Coatings for implementation by 31st December 1999. A further staged reduction has been proposed with a target date for compliance in line with European timescale. Water based alternatives prefered as they reduce exposure to solvents. It should not be forgotten that varnishes have a relatively high embodied energy in relation to mass and that they have a relatively short life compared to other construction elements. Durability assessment depends upon correct application in appropriate weather conditions. Construction phase score 2 if excessive waste paint is disposed of incorrectly.	5. Water based (max VOC levels: <140g/l) 6. Solvent based (max VOC levels: <600g/l)	Performance dependent on selection of specific products for application – confirm with manufacturer.	
	Multi coloured including special effect coatings	For example: Woodgrain effect	
	7. Water based (max VOC levels: <150gl/l) 8. Solvent based (max VOC levels: 350g/l)	Similar performance to woodstains and varnishes.	

Health Comment	Rank	Environmental Issues	Rank	Cost Rank
No significant risks foreseen unless fungicides or insecticides present (see application sheet 7.7). Painters/DIY should ensure adequate levels of ventilation. The remaining comments are to do with the resin base of the various products:	-			
		Pollution from manufacture. High embodied energy. Life expectancy score 2 for external applications.	1/1/1/0	100*
Polyurethane resins: no significant risk foreseen to occupant from fully cured resin. The process of resurfacing without precautions, will expose the DIY-er to inhalable dust and fume and skin contact with the resin. Skin and respiratory irritation and sensitisation may result. The resulting cough, breathlessness and wheeze may be a transient nuisance, or may recur intermittently long after cessation of exposure, rendering the activity unacceptable (3/3).	0/0	Pollution from manufacture. High embodied energy. Life expectancy score 2 for external applications. High level of solvents (VOCS)	1/2/1/0	192*
		Pollution from manufacture. High embodied energy. Life expectancy score 2 for external applications.	1/1/1/0	124*
Water based acrylic resins: no significant risk foreseen to occupant from cured resin. The DIY-er requires to avoid dust inhalation when sanding, and skin contact when resurfacing.	0/0	Pollution from manufacture. High embodied energy. Life expectancy score 2 for external applications. High level of solvents (VOCS)	1/2/1/0	162*
White spirit based alkyl resins: no significant risk foreseen to occupant from cured resin. The DIY-er requires to avoid dust inhalation when sanding, and skin contact when resurfacing.	0/0	Pollution from manufacture. High embodied energy. Life expectancy score 2 for external applications.	1/1/1/0	133*
		Pollution from manufacture. High embodied energy. Life expectancy score 2 for external applications. High level of solvents.	1/2/1/0	187*
		Pollution from manufacture. High embodied energy. Life expectancy score 2 for external applications.	1/1/1/0	151*
		Pollution from manufacture. High embodied energy. Life expectancy score 2 for external applications. High level of solvents.	1/2/1/0	169*

* supply only

Application 7.11 WALLPAPER AND WALLPAPER PASTES	Alternatives	Technical Comment	Rank
Typical Situation Means of providing decorative finishes to walls.	**Wallpapers** 1 Vinyl wall covering	Decorative covering primarily used where there is high water vapour levels and/or surface needs regular cleaning, eg. Kitchens.	
Technical Requirements To provide aesthetically pleasing finish and to cover minor surface defects. In some locations ability to withstand dampness and cleaning is an advantage. Pastes must be of adequate strength to retain paper until stripping.	2 Decorative wall papers	Decorative coverings which hide some surface defects. Not resistant to excessive levels of water vapours or cleaning.	
Decay and Degradation Factors Dampness, fungal attack, pollutants in indoor air (eg. tobacco smoke). Normal wear and tear. U.V. light. Maintenance work.	3 Decorative wall papers (recycled paper)	Decorative coverings which hide some surface defects. Not resistant to excessive levels of water vapours or cleaning.	
Guidance Notes System selection should consider the risk of dampness as a significant cause of decay and degradation in certain rooms or locations. Option 1 offers better performance in such circumstances at a slightly higher environmental impact. Otherwise selection can be made on aesthetic criteria. Perception of appearance is also the most significant factor in determining the life-expectancy of wall-papers. Good quality wallpapers can have a surprisingly long life in pollution free environments. Modern wall papers are innocuous in health terms. Historically arsenic pigments were thought to have been toxic when attacked by fungi. Modern fungicides should conform to health requirements in order to protect paper hangers and occupants of papered rooms. Historic materials could be toxic. So care is required for either removal or remedial work.	4 Wood chip	Wood chips encapsulated in paper. Surface needs painting.	
	5 Textured papers	Hides greater wall surface defects than ordinary wall papers. Usually needs decoration.	
	Wallpaper pastes 6 With fungicides	Used where there is possibility of fungi growth in damp conditions.	
	7 Without fungicides	Will not combat fungi growth in damp conditions.	
	8 PVA paste	Used for heavier wall papers.	

Health Comment	Rank	Environmental Issues	Rank	Cost Rank
No significant risk foreseen to occupants.	0/0	High embodied energy. Problems from pollution during manufacture but situation improving. Problems at disposal. Downstream 2 if disposed of by combustion.	2/0/1/1	241*
No significant risk foreseen to occupants.	0/0	Paper product with some toxic materials produced during manufacture. Biodegradable.	1/0/1/0	321*
No significant risk foreseen to occupants.	0/0	Less embodied energy and 60% less water consumption during manufacture. Biodegradable.	0/0/1/0	465*
No significant risk foreseen to occupants.	0/0	Paper product with some toxic materials produced during manufacture. Biodegradable.	1/0/1/0	100*
No significant risk foreseen to occupants.	0/0	Assumes paper product but upstream and downstream score increase by one if vinyl (see 1 above).	1/0/1/0	201*
See guidance notes.	0/0	Hazards from fungicide effects, disposal and upstream score.	1/0/1/1	166*
See guidance notes.	0/0	Usually mixture of methyl-cellulose and water – some are cold water starch.	0/0/1/0	121*
See guidance notes.	0/0	High embodied energy. Problems from pollution during manufacture. Problems at disposal.	2/0/1/1	100*

* supply only

Application 7.12 MAN MADE TIMBER BOARDS	Alternatives	Technical Comment	Rank
Typical Situation Sheet material used for flooring, wall lining, ceilings, cupboard carcass, worktop, shelving. May have a variety of decorative or moulded finishes/forms.	I. Plywood	Strength related to number of plys for given thickness. Exterior grades must be used in damp conditions. Contains organic adhesives; either urea formaldehyde, phenol, resorcinol or melanine formaldehyde.	
Technical Requirements Adequate strength for fixing and span over support system. Various alternatives have a wide range of properties.	2. Blockwood	Strength related to number of plys for given thickness. Exterior grades must be used in damp conditions. Contains organic adhesives; either urea formaldehyde, phenol, resorcinol or melanine formaldehyde.	
Decay and Degradation Factors Water, fungal attack, impact, abrasion, fire, frost.	3.Particle-board: a) Chip-board b) Cement bonded wood particle board	(a) Strength related to adhesive, size of chips and final density. Water resistant grades are superior in damp conditions. Adhesives; urea formaldehyde, phenol formaldehyde, isocyanurate based. Many have silicone water repellant. (b) More difficult to fix – needs to be drilled. Bonding agent is portland cement.	
Guidance Notes These materials find a wide range of applications some of which are explained in other data sheets. Performance characteristics depend upon the nature of the fibres used and on the type and quality of adhesives (alternatives 1, 2, 3, 6, 7, 8) or on pressure pressing (alternatives 4 and 5). Water resistant adhesives (usually resin based) provide better assurance against loss of strength due to dampness and/or in some materials (eg. chipboard) water resistance grades. In maintenance of dry conditions will ensure resistance to fungal attack but there is always a slight risk due to plumbing or roof leaks.	4. Hardboard	Can be used with a variety of surface finishes. Thinner sheeting requires closer spanned supports. May need conditioning. No adhesive used in manufacture. May have water repellant agent such as bitumen or silicone.	
The key issues from an environmental perspective are the source of the fibres/or timber plys and whether formaldehyde based resin adhesives are used.	5. Softboard	Normally used as pinboard or for sound or thermal insulation. No adhesive required. May have water repellant agent such as bitumen or silicone.	
Some block boards and plywoods are made using tropical hardwoods of a quality and durability which is usually unnecessary in the interior of buildings. These boards should be avoided unless it can be shown that the hardwood comes from sustainably managed sources. A number of alternatives (eg. chipboard) use a very high proportion of waste timber (either from timber processing or retrieved from construction waste) which would otherwise go to landfill. Formaldehyde resins have largely been reduced or replaced in the majority of chipboards available ameliorating the problems with off-gassing to the interior. Strawboard provides an application for organic material but is more susceptible to dampness than some of the alternatives.	6. MDF (Medium Density Fibre-board)	Mainly used in decorative products/applications. Organic resin content up to double that used in chipboard	
Alternatives 4 and 5 rely on pressing of fibres to provide mechanical strength in a similar way to paper making and in some cases use a proportion of recycled wood fibre. With all wood fibre products control measures are necessary to prevent inhalation of large quantities of wood dust from sawing and sanding.	7. Wood wool/ Cement board	More difficult to fix – needs to be drilled. Bonding agent is portland cement.	
	8. Straw boards	Used as a partition system. Decays if exposed to damp conditions. Resin based adhesives can be used.	

Health Comment	Rank	Environmental Issues	Rank	Cost Rank
Exposure to substantial amounts of fine wood dust is associated with adverse effects on breathing and with cancer, but would only be of concern in DIY sanding where the risk may be elevated to 0/3.	0/1	Impacts from Formaldehyde adhesives. Upstream score 1 for birch ply sourced in Europe. Downstream score 2 if burned at end of life. Reuse of sheet material difficult.	2/0/0/1	600*
Exposure to substantial amounts of fine wood dust is associated with adverse effects on breathing and with cancer, but would only be of concern in DIY sanding where the risk may be elevated to 0/3.	0/1	Impacts from Formaldehyde adhesives. Upstream score 1 for birch ply sourced in Europe. Downstream score 2 if burned at end of life. Reuse of sheet material difficult.	2/0/0/1	590*
(a) The state of the resin and the amount of ventilation will determine where on the scale 0/1 to 3/3 the application falls.	0/1	a) Impacts from Formaldehyde adhesives. Downstream score 2 if burned at end of life. Reuse of sheet material difficult.	1/0/1/1	280*
(b) No significant risk foreseen to occupants.	0/0	b) Impacts of cement manufacture.	1/0/1/0	376*
No significant risk foreseen to occupants.	0/0	Assume fibre from softwood and adhesive free manufacture. Upstream score 1 if not the case.	0/0/1/0	176*
No significant risk foreseen to occupants.	0/0	Assume fibre from softwood and adhesive free manufacture. Upstream score 1 if not the case.	0/0/1/0	100*
No significant risk foreseen to occupants. Exposure to substantial amounts of fine wood dust is associated with adverse effects on breathing and with cancer, but would only be of concern in DIY sanding where the risk may be elevated to 0/3.	0/0	Impacts from Formaldehyde adhesives. Downstream score 2 if burned at end of life. Reuse of sheet material difficult.	2/0/1/1	421*
No significant risk foreseen to occupants.	0/0	Impacts of cement manufacture.	1/0/1/0	440*
No significant risk foreseen to occupants. Is resin based adhesive used then the state of the resin and the amount of ventilation will determine where on the scale 0/1 to 3/3 the application falls.	0/0	If resin bonder used, upstream score 1. Often a composite with other facings, eg. Ply, complicates at disposal.	0/0/1/1	432*

* supply only

Application 7.13 DAMP PROOF MEMBRANES (DPM)/DAMP PROOF COURSES (DPC)/ DAMP PROOFING SYSTEMS	Alternatives	Technical Comment	Rank
Typical Situation Damp proofing in ground floors, basements and in walls.	1 Poly Vinyl Chloride (PVC)	Effective and common means of providing DPMand DPC. Needs effective lap at joints.	
	2 Asphalt	Often used for remedial treatment on existing floors.	
	3 Low Density Polyethylene (LDPE)	Effective and common means of providing DPM. Needs effective lap at joints.	
Technical Requirements Effective means of preventing dampness rising from ground or penetrating around doors and windows into the building interior.	4 Bitumen liquid applied (DPM only)	Requires protection by screed.	
Decay and Degradation Factors Frost, vermin, movement. Atmosphere pollution, chemical attack.	5 Pitch polymer	Effective means for DPC but sheet sizes available for DPM means lap joints required.	
Guidance Notes There are three main applications, DPC in walls and around openings, horizontal membranes in ground floors and tanking to walls and floors below ground level. Each of the physical barriers (alternatives 1-6) offer adequate performance in one or more of these applications. With sheet materials, good detailing and workmanship at joints is important to ensure effective damp-proofing performance. This is most critical in the more extreme dampness exposure position of basement tanking. With 'liquid' applied membranes (options 2 and 4) the construction of wall and floors must prevent structural movements which may disrupt or sever the membrane. In tanking all systems require effective detailing where service pipes and wires penetrate the membrane.	6 Slates (DPC only)	Traditional DPC in walls. Two course work – minimum requirement.	
	7 Silicone injection	For repair/improvement of old walls.	
	8 Electro-osmosis	For repair/improvement of old walls.	
Options 7, 8 and 9 are primarily intended for retro fit application in older buildings to remedy or alleviate dampness problems. Except for asphalt (alternative 2) the quantity of material in membranes is small so there is little to choose between alternatives in terms of environmental impact, although concerns about PVC may cause some to avoid this material.	9 Ceramic tubes	For repair/improvement of old walls.	

Health Comment	Rank	Environmental Issues	Rank	Cost Rank
No significant risk foreseen to occupants.	0/0	Problems with pollution from manufacture but situation is improving. Problems at disposal.	2/0/0/2	Refer to specialist suppliers
No significant risk foreseen to occupants.	0/0	Main impacts from bitumen – manufacturing pollution and medium embodied energy.	2/1/0/1	
No significant risk foreseen to occupants.	0/0	Moderate embodied energy. Pollution from petrochemicals.	1/0/0/1	
No significant risk foreseen to occupants.	0/0	Main impacts from bitumen – manufacturing pollution and medium embodied energy.	1/1/0/1	
No significant risk foreseen to occupants.	0/0	Moderate embodied energy. Pollution from petrochemicals.	1/0/0/1	
No significant risk foreseen to occupants.	0/0	Despolation from quarrying and quarry waste.	1/0/0/0	
No significant risk foreseen to occupants.	0/0	Solvents. Energy and materials involved in drilling processes. Silicone embodied in masonry therefore lost at disposal.	1/1/1/-	
No significant risk foreseen to occupants.	0/0	Moderate embodied energy in copper electrodes. Assume copper recycled at end of life.	1/0/1/0	
No significant risk foreseen to occupants.	0/0	Embodied energy of tubes. Energy and high use of drill bits involved in drilling processes.	1/1/1/0	

Application 7.14 BUILDING MEMBRANES	Alternatives	Technical Comment	Rank
Typical Situation Vapour control layers: breathing membranes Vapour checks Vapour barriers Sacking/Underfelt	1 Breathing Membranes (a) paper (b) poly-propylene	 Tears easily unless reinforced with additional fibres (eg. polyester). Tear resistant – higher strength.	 3 1
Technical Requirements Vapour control layers in general to control the passage of draught, penetrating water, moisture-vapour, dusts and fibres. Breathing membranes used on the cold side of the insulation preventing water penetration but permitting moisture vapour to escape. Vapour checks are used on the warm side of the insulate reducing moisture vapour penetration from the inside of the building. Vapour barriers are used in areas of higher temperature and humidity such as kitchens and bathrooms. Sarking/Underfelts used to restrict the passage of draughts and penetrating water as a back up system to tiles and slates in pitch roofs.	2 Vapour checks and barriers (a) Paper and aluminium foil (b) Poly-ethelene (c) Poly-propylene (d) Aluminium foil	 Requires stapling/fixing in place. Requires stapling/fixing in place. Requires stapling/fixing in place. Usually bonded to plasterboard.	 1 1 1 1
Decay and Degradation Factors UV radiation where exposed. Fire, vermin, birds, and tear impact particularly during construction phase. Damage due to maintenance and alterations.			
Guidance Notes A key issue from a technical standpoint is the degree of integrity and therefore airtightness achieved in the provision of vapour barriers. Airtightness is important to achieving full vapour control performance (but is not necessary for sarking felting applications). Airtightness can be improve by sealing joints, avoiding damage to the integrity of the membrane through good workmanship by avoiding cutting for routing service pipes and wires and sealing penetration of services through the membrane. Improved airtightness has implications for the health of occupants and so it is important that the minimum levels of ventilation are installed as required in building regulations (eg. fixed ventilation openings in windows, extractor fans in kitchens, etc). The concept of the breathing wall using porous materials and dynamic insulation is subject to ongoing research. As the quantity of material in membranes is small there is little to choose between alternatives from an environmental perspective, although concerns about PVC may cause some to avoid this material.	3. Sarking/ Underfelts (a) Bitumen felts (b) PVC (c) Poly-ethylene (d) Poly-propylene with polyester filler	 Natural fibre based systems. Tears relatively easily. Gradual embrittlement over time. More tear resistant – higher strength than 3(a). More tear resistant – higher strength than 3(a). More tear resistant – higher strength than 3(a).	 2 1 1 1

Health Comment	Rank	Environmental Issues	Rank	Cost Rank
No significant risk foreseen to occupants.	0/0			
		Natural material. Assumes high level of recycled material otherwise upstream score 1.	0/0/0/0	206*
		Moderate embodied energy. Pollution from petrochemicals. Downstream score assumes some recycling.	1/0/0/1	100*
No significant risk foreseen to occupants.	0/0			
		(a) Embodied energy of aluminium.	1/0/0/1	N/A
		(b) Moderate embodied energy. Pollution from petrochemicals. Downstream score assumes some recycling.	1/0/0/1	102*
		(c) Moderate embodied energy. Pollution from petrochemicals. Downstream score assumes some recycling.	1/0/0/1	102*
		(d) Embodied energy of aluminium. Impossible to recover aluminium at disposal.	1/0/0/1	N/A
No significant risk foreseen to occupants.	0/0			
		(a) Main environmental impact from bitumen. Manufacturing pollution and embodied energy.	1/0/0/1	187*
		(b) Moderate embodied energy. Problems from pollution during manufacture but situation is improving.	2/0/0/1	392*
		(c) Moderate embodied energy. Pollution from petrochemicals. Downstream score assumes some recycling.	1/0/0/1	171*
		(d) Moderate embodied energy. Pollution from petrochemicals. Downstream score assumes some recycling.	1/0/0/1	497*

* supply only

Application 7.15 GLASS AND OTHER GLAZING MATERIALS	Alternatives	Technical Comment	Rank
Typical Situation Transparent sheet material normaly applied in frames or other fixing system for windows, doors and roof lights.	**Glass** 1(a) Annealed	'Ordinary' glass used in the majority of domestic applications.	
Technical Requirements Translucent material to admit light and solar energy that is easily cleaned. Overall performance of glazing is very dependent upon the framing system. Selection will be determined by thermal, acoustic, safety, privacy and security requirements.	1(b) Laminated	Provide resistance to penetration by persons (unauthorised or accidental). Maintains its integrity. Used for safety and security.	
Decay and Degradation Factors U.V. radiation. Frost. Impact. Fire. Wear and tear. Movement. Long term durability is substantially influenced by the performance of the framing and fixing systems.	1(c) Toughened	Used for safety and strength characteristics. 4 times stronger than 1. Resistance to breakage.	
Guidance Notes Most of the comments assume that double glazing will be used in external windows. Single glazing is now mainly used for internal applications.	1(d) Wired	Fire resistance properties. Safety versions available.	
From a technical perspective double glazing with low E coating forms a minimum specification to reduce heat loss from housing. However, double glazing units have low life expectancy of around 10 years, due to the performance of the edge seals, when compared with the simple glazing of normal unsealed glass which can last the life of the building. Currently double glazed units tend to be discarded at the end of their life. To add to the energy savings resulting in the use of double glazing over single glazing, the short life of double glazing means that the material in the units should be recycled when the seals fail in order to justify the energy investment made in the original manufacture of the glass.	1(e) Textured/ Obscure	Decorative or obscuring for privacy.	
Plastic materials are gradually degraded by U.V. radiation and are more susceptible to scratching and abrasion.	**Glass and coatings** 2(a) Low E glass	Used on 1, 2 and 3 above. Improved thermal insulation properties.	
The main health risk is due to breakage. Regulations require toughened or laminated glass in high risk locations and applications, eg. in glazed doors.	2(b) Solar control glass (tinted)	Used on 1, 2 and 3 above. Reduces solar gains by reflecting sun light.	
Given the decision to use double glazing on the grounds of comfort and energy efficiency there is little to choose between alternatives from the point of view of environmental impact, except that the improved thermal performance of E glass will result in better return on investment in lower energy consumption in the heating season.	**Plastic** 3(a) Acrylic (Polymethyl metha-oxylate)	Robust material, but prone to scratching and discolourations.	
Solar control coatings are one way of reducing solar gains in summer through south facing glazing and can reduce or eliminate the need for mechanical cooling and the associated energy requirement in summer. Physical shading is an alternative to solar control coatings.	3(b) Poly-carbonate	Robust material, but prone to scratching and discolourations	

Health Comment	Rank	Environmental Issues	Rank	Cost Rank
No significant hazard foreseen in normal use.	0/0	*The upper ranking is for single glazing and the lower for double-glazing. High embodied energy, uses 20% recycled glass. Downstream score '0' if recycled or reused.	* 2/0/2/1 2/0/0/1	100*
No significant hazard foreseen in normal use.	0/0	High embodied energy, uses 20% recycled glass. Downstream score '0' if recycled or reused.	2/0/2/1 2/0/0/1	N/A
No significant hazard foreseen in normal use.	0/0	High embodied energy, uses 20% recycled glass. Downstream score '0' if recycled or reused.	2/0/2/1 2/0/0/1	150*
No significant hazard foreseen in normal use.	0/0	High embodied energy, uses 20% recycled glass. Downstream score '0' if recycled or reused.	2/0/2/1 2/0/0/1	163*
No significant hazard foreseen in normal use.	0/0	High embodied energy, uses 20% recycled glass. Downstream score '0' if recycled or reused.	2/0/2/1 2/0/0/1	100*
No significant hazard foreseen in normal use.	0/0	High embodied energy, uses 20% recycled glass. Coatings may mean glass is difficult to recycle.	2/0/2/1 2/0/0/1	125*
No significant hazard foreseen in normal use.	0/0	High embodied energy, uses 20% recycled glass. Coatings may mean glass is difficult to recycle.	2/0/2/1 2/0/0/1	113*
No significant hazard foreseen in normal use.	0/0	Moderate embodied energy, pollution from petrochemicals. Downstream score '0' if recycled.	2/0/2/1 2/0/2/1	N/A
No significant hazard foreseen in normal use.	0/0	Moderate embodied energy, pollution from petrochemicals. Downstream score '0' if recycled.	2/0/2/1 2/0/2/1	N/A

* supply only

CHAPTER SEVEN

Hazardous Materials in Existing Buildings

M. ANDERSON
N FORD, BSc
C G MARCH, BSc (Tech) MCIOB

7.1 INTRODUCTION

Applications of hazardous materials which, although no longer available or permitted, can still be found in existing buildings have been included in the comparisons made in the application sheets in Chapter 6. The purpose of this chapter is to expand on this and provide general information, comment and advice on whether or not to remove existing hazardous materials, and if so, how to carry out this operation and eventually dispose of the offending material in a save manner. The prime causes for concern are the majority of asbestos applications and lead used in water supply installations and paintwork.

The procedures are given only in outline, as this chapter is supplementary to the main theme of the text dealing with the selection of materials. Also individual local authorities have their own requirements and codes of practices.

7.2 ASBESTOS

7.2.1 When should asbestos be removed?

There are obvious cases where asbestos should be removed because there is an established risk to the health of the occupants of the building. For example, circumstances where the fibre if being released in to the atmosphere due to degradation of the material by abrasion, decay and similar factors and where no reasonable amount of repair will stop this from occurring. On the other hand, materials containing asbestos fibre are used in applications where the material is completely contained, for example when encased inside a double-walled stainless steel flue, where the likelihood of the fibre being released in normal use is virtually nil.

So, the question arises, should the material be removed or left alone if it is not subject to damage, or if this has occurred and further damage is not expected, should it be repaired or replace with an alternative material? This is the problem for the building owner, manager, designer, surveyor or contractor concerned to address. It is probable that at the time of writing the building owner and landlord of residential property will eventually be made responsible as, at the time of writing (2000), the consultative document circulated by the HSE is recommending that non-residential building owners should have this responsibility placed on them.

Some would argue that any asbestos found should be removed, based upon the extreme view that 'it only takes one asbestos fibre to cause asbestos related diseases', but others would take a more liberal view in their interpretation of risk. One should not underrate the sometimes enormous costs of removal or the difficulty in complete removal of the material which can lodge in voids and cavities. However to take any short cuts almost certainly means that irresponsible or incorrect removal will take place, which will be hazardous to both operatives and occupants. If on the other hand a decision is made to stabilise and/or contain the asbestos in situ, it ultimately leaves the problem of safe demolition. This process may not be as well controlled as phased removal. It is therefore essential to ensure that all of these processes are properly pre-planned and carried out by a responsible contractor. Due to the complexity of the issues it is impossible to make this judgement on behalf of others in isolation and therefore decisions in each individual case must be taken on merit.

7.2.2 Identification and sampling of asbestos

The most likely places where asbestos may be found in low-rise residential buildings have been identified in the application sheets in Chapter 7. It should be noted that over the lifetime of the building there is the possibility that asbestos fibre may have been released due to previous alteration and decoration. Also consideration must be given to the original installation where inappropriate equipment may have been used when cleaning up on completion, thus causing pollution. Therefore, fibres may have spread into other locations other than where originally installed.

However, asbestos material may have been painted over or otherwise concealed, thus disguising its true nature and, even if this is not the case, many of the substitute materials may have similar appearance even to the experienced eye. Even if it has been established that the material contains asbestos fibre, it will probably not be known if the fibre is white, brown or blue (see chapter 2). To be sure, bulk and air samples will have to be taken by an United Kingdom Accredited Service (UKAS) accredited person and taken away for analysis. The Environmental Health Department of the local authority should be contacted if there is any doubt at all, they will provide, proper advice on, a sampling and analysis service. Where large quantities are involved it might be prudent to consult a specialist consultant who, besides providing the sampling and analysis, service would specify the method and supervise the removal of the offending material. The following notes provide only an outline of the general procedure. Readers *must not* attempt to carry out sampling and removal procedures without seeking proper advice from the Local Authority Health Department, Health and Safety Executive, suitably qualified laboratory or consultant.

Advice on the procedures and precautions to be taken are outlined in the Health and Safety Executive Code of Practice: Control of Asbestos at Work Regulations (1987) 3rd Edition and relevant guidance notes.

The vast majority of applications in domestic buildings will involve either asbestos cement and/or insulation board. The key points are outlined below.

- Whenever it is necessary to work on or disturb material containing asbestos fibre an analysis of its composition and density of the fibre should be first carried out to establish whether or not it is necessary to use a licensed or unlicensed contractor to carry out the work. The presence of asbestos may be indicated on original building plans or specifications. Information about the presence of asbestos may also be available from the architect or builder who constructed the building, or from the original supplier of the insulating board or cement product if known. A voluntary labelling system was introduced for asbestos products in 1976 using a logo to indicate the presence of asbestos. However, very little asbestos containing materials found in existing buildings is likely to bear this label.

- If in doubt, the only satisfactory way of determining whether or not asbestos is present in cement is by bulk sampling and laboratory analysis. However even the sampling operation can put people at risk so it should only be done when the alternatives above have been tried and when there is a specific need to confirm the presence of asbestos. Sampling should only be carried out by someone who is UKAS accredited. Once asbestos has been identified, records should be made and kept available for any future work activity, Control of Design and Management Regulations (CDM) requirements need also to be considered as it is always good practice to hold an asbestos audit file for all buildings. Whilst asbestos is just another occupational hazard, it is the worst kind with regard to statistics.

- As insulation boards and cement products will normally be of uniform composition, there should in most cases be little difficulty in selecting a site for sampling which is not only representative, but also accessible and, importantly, can be easily cleaned and repaired after sampling. Asbestos insulating board may, however, have been repaired or extended with non-asbestos materials. It is therefore important to examine all materials for changes in characteristics or modifications/repairs, which may indicate a different composition and ensure that the samples are taken of all types of materials present. It is not sufficient merely to sample dust deposits in the vicinity of the material. Removal of samples must not compromise any fire resisting properties of the structure.

■ Sampling techniques used should minimise the release of fibre and cause minimum disturbance to the rest of the installation. This is particularly important when working overhead. The sampler must be UKAS accredited. Air sampling may also be necessary. Where old asbestos is involved, it is important to confirm the fibre type by sampling and analysis so that appropriate precautions can be taken.

■ The Health and Safety Executive (HSE) must be given 14 days notice before any work can commence on insulating board, coatings or insulation material. If there is any doubt of the type of asbestos, then it must be assumed that it is either blue of brown and the HSE notified. It should be noted that the licensed contractor must give 14 days notice to the HSE, irrespective of the fibre type, as a condition of their licence.

7.2.3 Repairing damage to asbestos cement and insulating board

Generally speaking, whilst it is possible with the techniques and procedures available, it is recommended that repairs are not undertaken, because normally the material is relatively old and may continue to break down. Also the possible added weight of the repair could cause extra stresses increasing the risk of further degradation.

Rigid asbestos materials such as insulating board or asbestos cement may be sealed by painting. The surface should be prepared using chemical cleaners and damaged areas repaired with a substitute material. Asbestos materials should not be sanded, wire-brushed or power washed. Dusty surfaces must only be cleaned with an industrial vacuum cleaner fitted with a high efficiency filter constructed to BS415 Appendix C, type H or wiped with a damp cloth which is disposed of whilst still damp in a sealed bag. Asbestos cement used externally may need treatment with a biocide to remove algae before painting.

Insulating board can be painted with a proprietary sealant specific for its use. Asbestos cement is alkaline and should be primed with an alkali-resistant primer or a chlorinated rubber or oleo-resinous paint followed by one or more topcoats. Where possible both sides should be painted. In any event it is important to refer to the product manufacturers specification.

Where a higher degree of protection from damage is required, a number of other sealing systems are available:

a) flexible or semi-flexible polymeric or bitumen coatings

b) inorganic cement type coatings

c) formed sheet or panels

The choice of sealing system depends on the nature of the asbestos material and its location, the degree of damage protection required and any surface flammability requirements. It is advised that users of any of these methods seek advice from the manufacturers or an accredited consultant.

Where asbestos insulation is being used for fire protection, it is important that the fire hazard is not increased by the use of combustible sealants. The sealed materials must meet the standards for spread of flame as specified in the Building Regulations. Normal paints may not achieve this standard, but specially formulated sealants are available.

Arrangements should be made to alert maintenance workers to the existence of asbestos material. In the case of housing, occupants should be made aware through adequate consultation of any asbestos material and advised of appropriate precautions and, if plans exist, the presence of asbestos should be recorded. Sealed asbestos should be checked by a qualified surveyor at regular intervals, at least once per year, to ensure that the sealing is intact.

7.2.4 Removal and disposal procedures

Removal is usually considered necessary for friable or damaged asbestos materials and in itself is a very hazardous procedure both to the occupants and the operatives. Unfortunately the restrictions and regulations placed on the operatives have to be so stringent for their own safety that there is the risk that they will take shortcuts, thus endangering their own health. Whilst not in the brief of this section of the text, it cannot be over stressed how important it is that operatives engaged in this work are properly supervised and controlled, not just for their own sake, but for the safety of others using the premise after they have finished work.

When it is necessary to remove asbestos containing materials, it is essential that a properly qualified contractor be employed. Properly qualified means experience as well as being licensed to carry out the work. Currently the problem is that possession of a licence is no guarantee the work will be executed in a proper fashion. It should be noted that it is a requirement to use a licensed contractor to remove friable insulating material and insulating board. A licensed contractor is more likely to be familiar with the problems of removal. It is not just a case of ensuring competent contractors are employed and supervised carrying out the removal, it is also important that the continuous monitoring by air sampling is carried out at the work place and surrounding areas.

When choosing a competent contractor it is recommended that the following facts be established:

that they

a) have the relevant experience

b) have adequate resources – both financial and equipment

c) have adequate insurance cover

d) carry out appropriate third party training and have an on going training programme for their own operatives

e) have quality systems in place

f) have no conditions on their licence preventing them from carrying out the specific work

g) have proof of similar type of work.

It would also be advisable to take up references.

It is important that all removal must be carried out in a controlled manner, which contains any material fibre release not only in the work zone, but also at the point of removal. A specific method and plan of work must be prepared for any job, however small. The method could well vary depending upon the problem encountered in each case. A 'Duty of Care' rests with the contractor as they are the experts, but this duty cannot be abdicated by the employer if he has not checked out the credibility of the contractor.

Equally, disposal procedures, both in transportation to the disposal site and when in place in the land fill site, must be controlled and carried out correctly or long-term environmental problems could ensue.

Any container used for the disposal of asbestos waste should be

a) made from impermeable material

b) strong enough to remain dust tight even under wet conditions and

c) be adequately labelled. The markings being durable and of a type which cannot be detached from the container.

When each container is filled

a) it should be sealed to prevent the escape of dust during handling, transportation and disposal

b) its external surface should be cleaned, and

c) it should be removed from the immediate working area to await removal to an authorised solid waste disposal site.

A system should be set up to ensure that the individual containers awaiting removal to an authorised land fill site are kept in a secure area set aside for storage.

Asbestos waste should be disposed of only

a) at a waste disposal site licensed for the purpose by the appropriate Environmental Agency and

b) accordance with the requirements of that authority

The local Environmental Health Officer and the Health and Safety Executive are able to provide the necessary local information and advice. The Metropolitan Boroughs and County Councils should be able to provide disposal facilities for domestic toxic waste and will give details of appropriate sites.

It will be obvious from reading the previous sections on the removal and disposal of asbestos containing materials that the procedures entail considerable expense when carried our correctly. However the consequences of not doing so could have profound effects both on the health of those involved in removal and the building occupants as well as long-term environmental impact.

Mike Anderson is Managing Director of LAR Ltd

7.3 LEAD

7.3.1 Water supply

Where lead pipework is known or suspected in existing buildings, the action to be taken is dependent upon the plumbosolvency of the water. Contact with lead during distribution does not necessarily result in contamination of water. The physical and chemical characteristics of the water – for example, acidity, hardness and temperature – determine its ability to dissolve lead. Soft acid water generally shows the greatest plumbosolvency, though some hard alkaline waters have been shown to be plumbosolvent.

The lead content of water from the tap depends upon the particular combination of circumstances prevailing before and drawing of a sample. Given the number of variables involved – for example, pipe length and condition, time for which the water has been standing and water flow rate – it is not surprising that the lead content of random samples taken at the same tap can vary considerably. Wide variations in the amount of water drunk and consumed indirectly by different people also make it difficult to estimate to what extent lead in tap water contributes to the average individual's total intake of lead, though it has been estimated that an adult's intake from tap water beverages and cooking water is usually about 10µg/day, and possibly 30 times this amount in an area where plumbosolvency is a problem. The main areas are the North of England, Wales and parts of Scotland, although this by no means identifies every area of concern. Designers should contact local water companies for detailed advice on water quality.

There are still households in Great Britain, which use water that has at some stage passed through lead pipework or tanks. Lead can be found in the distribution system, in the domestic service branch pipes and also in pipework and storage within the home, although by this date it is thought that the majority of lead applications have now been replaced by necessary maintenance and improvement work.

Identification may pose a problem. Surface mounted pipe is obvious, but real difficulties arise in trying to establish the nature of alterations during the life of the building. For example, what appears to be a copper supply pipe may simply be a short length of copper jointed to an otherwise lead pipe and subsequently concealed by a new floor slab. It would therefore be advisable for designers to investigate fully either by instigating an analysis of water quality or by excavation of existing pipework when alterations or refurbishment is contemplated. The local environmental health office can advise on the address of the public analysts or appropriate private laboratory that can carry out the analysis. It is also worth while establishing whether grants are available for replacement work.

It should be noted that removal of lead water supply pipes in the house might not totally resolve the problem since the communication pipe, owned and serviced by the water company, may also be of lead, although by this date this is increasingly unlikely.

While replacement is the most positive method of removing the risks posed by pipework, it may not always be necessary to do this immediately as water companies are investigating the effectiveness of treating water supplies to reduce plumbosolvency in high risk areas. In such circumstances replacement might be delayed until a convenient moment such as when a major overhaul of the plumbing system is required, when refurbishment of the property is in hand or when the property changes ownership. The water company should be contacted for advice.

When replacement is contemplated advice on alternative materials is provided in application sheets 5.2 and 5.4 in chapter 6. The safest method of disposal of lead is to recycle.

7.3.2 Paint

Chemical analysis is required before one can say with any certainty whether or not a particular piece of paintwork has a significant lead content. However, history gives us a useful general guide. Before the First World War, lead based paint was used extensively on walls, wood and metals, both indoors and outdoors. All paintwork and priming thought to date before this is likely to contain significant amounts of lead. Technical changes after about 1920 caused steady reductions in both the average lead content of leaded paints and their use, first for indoor work, and later even for outdoor priming. So leaded paintwork is likely to be found on the exteriors of inter-war buildings and is not uncommon on interior surfaces, especially in the priming coats. The overall reduction in lead content has continued and by 1987 was effectively eliminated in all domestic paints. Thus there is still the possibility that unsuitable leaded paint or primer will have to be removed as it might be disturbed during redecoration.

A small flake of paint is all that is required for an inexpensive test to establish the lead content. It must be borne in mind that recent unleaded paint may conceal older lead based paint of primer and that older woodwork may have some leaded primer left on it after it has been stripped. Exterior metal work of all but the newest is very likely to have at least a lead based primer on it, unless it has been stripped recently and painted with lead free paint and primer.

Leaded paint only becomes a serious hazard when it is disturbed and only then if the disturbance is such that people breathe or eat dust or debris. Sound paintwork, free from peeling, cracking, chipping or other deterioration poses no problem except in the case of children suffering from pica (pica is the habit of some infants to chew parts of buildings).

Although children with pica do not always chew paint of painted surfaces, where a child has this condition the only completely reliable safety measure is to remove the child to a place which is known to be free from leaded paint, especially on accessible surfaces. Failing that, the child should be closely supervised and all accessible surfaces in the home should be stripped with care and repainted with lead free paint, taking care to follow the instructions below.

Once lead paintwork is detected it does not generally follow that it needs to be removed. Wholesale removal is generally neither feasible nor cost effective and, if not done with scrupulous care, will probably maker matters worse by releasing lead dust and particles around the home to be inhaled or ingested. When paintwork is sound, and children are not exposed to it, it is unlikely to present any serious hazard, whatever the lead content, if it is left alone or simply covered with modern paint in the course of normal redecoration. Preparation of sound paintwork should be restricted to cleaning using ordinary domestic cleaning solutions and very light abrasions using 'wet and dry' paper in order to scratch the surface and give a key for the following coats. However, there may eventually come a time when the paint will of necessity need to be stripped off due to decay or at major refurbishment.

Where leaded paint is flaking or crumbling, or needs to be removed for some other reason, there are a number of precautions, which should be taken by both professional and amateur decorators. Dry sanding, whether for surface preparation or complete removal of the paint is hazardous, particularly when power tools are used. The large quantities of dust released may be directly breathed in and contribute to high indoor lead concentrations for extended periods.

The hazards fully justify the advice that dry sanding should never be performed indoors for the removal of lead paint. Burning or other stripping methods using heat can generate lead-rich fumes, which can be dangerous to the decorator and occupant if exposed for long periods. Hot air tools are now widely available, which soften the paint without generating fumes, providing the temperature is below 500°C, and allows it to be scraped off. However it is quite easy to overheat the paint with these hot air machines so they must be used with care. There is also the danger of generating lead rich dust as the machine will blow particles or flakes of paint about as it is being scraped off, causing widespread dust contamination.

Wet sanding avoids the hazards of both dry sanding and burning, but is slow for large areas and messy for domestic purposes. Chemical paint strippers will give satisfactory results, but they are expensive, caustic to the skin and give off fumes, which are hazardous if breathed to excess.

One must conclude that there is no completely 'safe' method of removing lead paints, but in overall terms chemical strippers appear to pose the least hazard, provided adequate ventilation can be assured. It is essential that DIY decorators provide adequate through ventilation. In order to achieve this it is usually necessary to open sufficient windows and doors so that air moves freely across the whole house while stripping and painting is in process and also until the paint has dried. Professional decorators should take adequate precautions to protect themselves from dust and fumes using appropriate breathing apparatus.

7.3.3 Dust

Dust is released by alterations and refurbishment of older property and by paint stripping. It is possible that the dust in the floor and roof voids together with service ducts may include raised levels of lead dust due to previous decoration of high lead content paints and general environment levels. Dust may also contain aspestos fibres. Care should be therefore exercised, particularly with refurbishment so that all dust is collected by vacuum equipment at each stage of the work and properly disposed of as below. Suitable vacuum cleaners are the HEPA types which have high efficiency filters.

7.3.4 Disposal

Lead pipe and sheet have significant scrap value and this will doubtless form one method of disposal in the short term, allowing recycling of the material, but this begs the question of the future. Local authorities will provide expert advice on the disposal of large quantities of waste.

Paint flakes and scrapings along with dust from refurbishment and alterations should be collected in stout sealed plastic bags and disposed of by an approved incineration facility.

Norma Ford is Senior Lecturer in the School of Environment and Life Sciences, University of Salford

7.4 CONCLUSION

The steps to be taken when a hazardous material is suspected in an existing building can be summarised as follows:

    Identification

    Measurement

    Assessment

    Sealing and/or

    Removal

    Disposal

Although these stages were originally outlined for asbestos, it should prove effective for other appropriate materials

The building owner, tenant, surveyor or architect will make initial identification of any hazardous material. The function of the building consultants after this phase will essentially revolve around managing the process and ensuring the best advice is available to the client. It is essential that reliable assistance and guidance is sought from the local authority environment health department, approved laboratory or specialist consultant.

Detailed measurements must be made, using appropriate techniques, to identify precisely the nature of the material concerned. Exploration and measurement of dust and fibres in wall, floor and roof voids, service ducts etc. may give a useful indication of the presence and type of material.

Assessment of the field data and any laboratory analysis must attempt to establish the likely exposure in order to come to an informed and realistic view of the appropriateness of removal, sealing or other measures and to balance the various risks involved. Sealing and or removal exercises require careful planning especially if the hazard is serious and extensive consultations with all parties involved is essential in this type of operation.

Finally, it is essential that a responsible attitude be taken over disposal of the material. If precautions and compliance with regulations are not observed, the problem is merely passed on to others and general environmental contamination may result.

The authors believe that all hazardous materials once identified, if not removed, should be recorded and this information passed on to prospective purchasers or tenants.

CHAPTER EIGHT

Appendix

As has been outlined previously, manufacturers marketing literature quite rightly concentrates on the positive functional and aesthetic attributes of their products. They are unlikely to wish to emphasise the negative issues. It is unsatisfactory where this promotional information is provided without the background context of the health and environmental information necessary for good decision making. Until such time as there is a real change of attitude and manufacturers provide detailed information as a matter of course either in their main literature or as an addendum, it will still be necessary to request this information separately. It is important that this information is collected not just to assist in the decision as to which material to select, but also to establish the cumulative amount or total load of a similar constituent from all the materials selected that is either completely contained or might be released over the life of the building.

With this information, a more informed judgement should be possible, not just in terms of the health of the occupants but also in terms of making, environmental assessments in the selection of materials.

Further it is anticipated that an abbreviated version of this text will be produced and made widely available for use by the general public to assist them when purchasing property. Professionals involved in the development design, construction and selling of residential buildings should therefore be more prepared to answer the kinds of questions on health and environmental issues that prospective purchasers may ask.

To assist in this, two forms have been included in this appendix.
8.1 Product Composition, to elicit information from manufacturers and is designed for two main reasons. Firstly to establish the constituent parts of the product and their potential health risk and secondly to obtain some information about the environmental implications in terms of how far the product (or component parts) has travelled and whether or not there is any recycled content.

8.2 Materials in existing and new buildings, for use in establishing the materials in existing or newly built residential dwellings at the point of sale. This would be useful for prospective buyers of properties in clarifying what potential hazards exist, and have been designed with this in mind. however it is of equal use in principle for the purchase of any property to establish a record of materials in the building to ensure that users of the building are not placed at risk especially when maintenance work is either being carried out or completed.

8.1 Product Composition

Product	Description:		Reference:	

Name of supplier:

Address: Telephone No:

       Fax No:

       E-mail address:

Main constituents	Approx.% of total product	Health and safety data sheets required (✓)	Country of origin or manufacture	Estimate % re-cycled content

Total embodied energy of the product:	Per volume	or Kg:	or Unit:

Potential to recycle all or part of product:

Any other relevant environmental and health data:

Date Requested: Date Received:

Please return to: Telephone No:

Name: Fax No:

Address: E-mail address:

8.2 Materials in existing and new buildings

To the vendor's solicitor

Please provide the following information about the materials the named building is constructed from. If unable to answer any question or the specific component is not part of the construction, please give reasons in the **comments** column.

If any alterations or replacement has occurred, please give approximate dates in the **comments** column along with a brief description of the work and composition of replacement materials. Use a separate piece of paper if necessary. The numbers given in the reference column refer to the application sheets in Hazardous Building Materials.

Address of Property	Developer/Estate Agent	Vendor's Solicitor	Purchaser's Solicitor
Number or name of property:	Name:	Name:	Name:
Name of Street/Road:	Street/Road:	Street/Road:	Street/Road:
Town/City:	Town/City:	Town/City:	Town/City:
Post Code:	Post Code:	Post Code:	Post Code:
Date of Survey:	Telephone No:	Telephone No:	Telephone No:

Date of initial construction of the property:

Main Element		Sub-Element	Yes/No	Materials of Construction	Comments	Application sheet reference
Foundation						
Basement	Q.1	Walls				7.1/7.2
	Q.2	Floor				7.1/7.2/7.3
	Q.3	Vertical Dpm				7.1/7.2
	Q.4	Horiz. Dpm				7.13
						7.13
Ground Floor						
Suspended Floor	Q.5	Floor				7.1/7.2/7.6/7.7
	Q.6	Insulation				2.1/2.4
	Q.7	Sleeper wall				7.3
Solid Floor	Q.8	Floor				7.1/7.2/
	Q.7	Insulation				2.1
	Q.9	DPM				7.13

Main Element		Sub-Element	Yes/No	Materials of Construction	Comments	Application sheet reference
Upper Floors						
Suspended Floor	Q.10	Floor				7.1/7.2/7.6/7.7
	Q.11	Insulation				2.1/2.4
Solid Floor	Q.12	Floor				7.1/7/2
	Q.13	Insulation				2.1
Walls (External)						
Masonry	Q14	Solid				7.3
	Q15	Cavity-inner leaf				7.3
	Q16	Cavity- Ext. leaf				7.3
Timber frame	Q17	Inner leaf				7.6/7.7
	Q18	External leaf				7.3/7.6/7.7
	Q19	Breather membrane				7.14
Insulation	Q20	External				2.3
	Q21	Cavity - full				2.2
	Q22	Cavity - partial				2.2
	Q23	Internal				2.4
	Q24	Timber frame				2.5
Walls (Internal)	Q25	Load Bearing				7.3
	Q26	Non-loadbearing				4.1
Roof - Pitched	Q27	Structure				7.6/7.7
	Q28	Roof covering				1.1/1.2
	Q29	Sarking felt				7.14
	Q30	Fascia, soffit etc				1.3
	Q31	Verge				1.4
	Q32	Flashings				1.5
	Q33	Gutters				1.6
	Q34	Downpipes				1.6
	Q35	Insulation				2.6
Roof - Flat	Q36	Structure				7.6/7.7
	Q37	Roof covering				1.2/1.7
	Q38	Fascia, soffit etc				1.3

182

Main Element		Sub-Element	Yes/No	Materials of Construction	Comments	Application sheet reference
Upper Floors						
	Q39	Verge				1.4
	Q40	Flashings				1.5
	Q41	Gutters				1.6
	Q42	Downpipes				1.6
	Q43	Insulation				2.7
Internal Wall	Q44	Kitchen				4.3
Finishings	Q45	Bathroom				4.3
(plaster, tiles etc)	Q46	Utility room				4.3
	Q47	Living rooms				4.2/4.3
	Q48	Bedrooms				4.2/4.3
Internal Wall	Q49	Circulation				
Finishings Cont.						
Internal Fl	Q50	Kitchen				4.5/4.6
Finishings	Q51	Bathroom				4.5/4.6
	Q52	Utility room				4.5/4.6
	Q53	Ground Floor				4.5/4.6
	Q54	First Floor				4.5/4.6
	Q55	Carpets				4.7
External Wall	Q56	Rendering				4.8
Finishings	Q57	Tiles/Planks				4.8
Services	Q58	Hot and cold water pipework				5.3
	Q59	Cold water storage tanks				5.4
	Q62	Boiler flue				5.6
	Q63	Chimney flue				5.6
	Q64	Mains water supply				5.3
Windows	Q65	Window				3.1
	Q66	Glazing				3.4/7.15
	Q67	Paint				7.9
	Q68	Putty				3.5

Main Element		Sub-Element	Yes/No	Materials of Construction	Comments	Application sheet reference
Doors	Q69	Sealants				3.5/7.5
	Q70	External				3.1/3.7
	Q71	Paint				7.9
	Q72	Sealants				3.5/7.5
	Q73	Internal				3.8
	Q74	Paint				7.9
Intern Joinery	Q75	Paint				7.9
	Q76	Architraves				4.4/7.6
	Q77	Skirtings				4.4/7.6
	Q78	Beading				4.4/7.6
External Works		Flat roof tiles				6.2
		Pathways				6.1
		Driveway				6.1
		Patio				6.1
		Boundary fencing and walling				6.3

I confirm that the information given in this form are correct.

Signature (a chartered professional) :   Dated:

CHAPTER NINE

Bibliography

Bell,R., Lowe, R. and Roberts, P. *Energy Efficiency in Housing.* Avebury. Aldershot. UK (1996)

Brandon, P. Bentivegna,V. and Lombardi, P. (Eds), *Evaluation in the Built Environment for Sustainability*, Chapman & Hall, London (1997)

Brandon, P. Sustainability in Management and Organisation; the Key Issues? *Building Research and Information* (1999) 27 (6) pp 391-397

Cooper, I. Which Focus for Building Assessment Methods – Environmental Performance or Sustainability? *Building Research and Information* 27 (4/5) pp321-331. E & F Spon, London (1999)

Curwell, S. and Cooper,I. The Implication of Urban Sustainability. *Building Research and Information* – special issue. Sustainability: an International Perspective. (1998) 26 (1) January/February pp17-28

Curwell, S.R., Hamilton,A, and Cooper,I. The BEQUEST Network:Towards Sustainable Urban Development, *Building Research and Information* 26 (1) January/ February pp56-65 (1998)

Curwell, S.R. and March,C.G. *Hazardous Building Materials :A Guide to the Selection of Alternatives*, E & F.N. Spon, London (1986)

Curwell, S.R., Fox, R.C. and March,C.G. *Use of CFCs in Building*, Fernsheer Ltd (1988)

Curwell, S.R., March, C.G. and Venables, R. *Buildings and Health*, RIBA Publications (1990)

Edwoods, B., *Green Buildings Pay.* E & FN Spon. London (1998)

Environmental Impact of Materials – Volume A : Summary, CIRIA Special Publication (1995)

Environmental Impact of Materials – Volume B: Mineral Products, CIRIA Special Publication (1995)

Environmental Impact of Materials – Volume C: Metals, CIRIA Special Publication (1995)

Environmental Impact of Materials – Volume D: Plastics and Elastomers, CIRIA Special Publication (1995)

Environmental Impact of Materials – Volume E :Timber and Timber Products, CIRIA Special Publication (1995)

Environmental Impact of Materials – Volume F: Paints and Coatings, Adhesives and Sealants, CIRIA Special Publication (1995)

The Good Wood Guide, Friends of the Earth (1990)

The Green Construction Handbook, JT Design &Build (1993)

Gribbin, J. *The Hole in the Sky: Man's Threat to the Ozone Layer.* Corgi, London (1988)

Hall, K. and Warm, P. *Greener Building: Products & Services Directory, 3rd Edition*,The Green Building Press (1995)

Howard, N. Sinclair, M.and Shiers, D. The Green Guide to Specification, Post Office Property Holdings (1996)

Jackson,T. *Materials Concerns: Pollution, Profit and Quality of Life.* Routlege, London. (1996)

Jenks,M., Burton, E and Williams, K. (Eds), *The Compact City: a Sustainable Urban Form?* E & FN Spon. London (1996)

Raw, G.J. and Hamilton, R.M. *Building Regulation and Health*, Building Research Establishment (1995)

Woolley T., Kimmins S., Harrison P. and Harrison R. *Green Building Handbook*, E & F.N. Spon (1997) (Second Edition 2000)

Yang, K. *The Green Skyscraper: the Basis for Designing Sustainable Intensive Buildings*. Prestel, Munich (1999)

Index